W0106851

RESIDUE REVIEWS

VOLUME 33

RESIDUE REVIEWS

Residues of Pesticides and Other
Foreign Chemicals in Foods and Feeds

RÜCKSTANDS-BERICHTE

Rückstände von Pestiziden und anderen
Fremdstoffen in Nahrungs- und Futtermitteln

Editor
FRANCIS A. GUNTHER

Assistant Editor
JANE DAVIES GUNTHER

Riverside, California

ADVISORY BOARD

F. Bär, Berlin, Germany • F. Bro-Rasmussen, Copenhagen, Denmark
J. W. Cook, Washington, D.C. • D. G. Crosby, Davis, California
S. Dormal-van den Bruel, Bruxelles, Belgium
C. L. Dunn, Wilmington, Delaware • H. Frehse, Leverkusen-Bayerwerk, Germany
J. C. Gage, Macclesfield, England • H. Geissbühler, Stein AG, Switzerland
S. A. Hall, Beltsville, Maryland • T. H. Harris, Bethesda, Maryland
L. W. Hazleton, Falls Church, Virginia • H. Hurtig, Ottawa, Canada
O. R. Klimmer, Bonn, Germany • G. K. Kohn, Richmond, California
H. F. Linskens, Nijmegen, The Netherlands • H. Maier-Bode, Bonn, Germany
N. N. Melnikov, Moscow, U.S.S.R. • R. Mestres, Montpellier, France
P. de Pietri-Tonelli, Milano, Italy • R. Truhaut, Paris, France

VOLUME 33

SPRINGER-VERLAG
BERLIN • HEIDELBERG • NEW YORK
1970

ISBN 978-1-4615-8469-8   ISBN 978-1-4615-8467-4 (eBook)
DOI 10.1007/978-1-4615-8467-4

All rights reserved. No part of this book may be translated or reproduced in
any form without written permission from Springer-Verlag.

© 1970 by Springer-Verlag New York Inc.
Softcover reprint of the hardcover 1st edition 1970
Library of Congress Catalog Card Number 62-18595.

The use of general descriptive names, trade names, trade marks, etc. in this
publication, even if the former are not especially identified, is not to be taken
as a sign that such names, as understood by the Trade Marks and Merchandise
Marks Act, may aeco dingly be used freely by anyone.

Title No. 7852

Preface

That residues of pesticide and other "foreign" chemicals in food-stuffs are of concern to everyone everywhere is amply attested by the reception accorded previous volumes of "Residue Reviews" and by the gratifying enthusiasm, sincerity, and efforts shown by all the individuals from whom manuscripts have been solicited. Despite much propaganda to the contrary, there can never be any serious question that pest-control chemicals and food-additive chemicals are essential to adequate food production, manufacture, marketing, and storage, yet without continuing surveillance and intelligent control some of those that persist in our foodstuffs could at times conceivably endanger the public health. Ensuring safety-in-use of these many chemicals is a dynamic challenge, for established ones are continually being displaced by newly developed ones more acceptable to food technologists, pharmacologists, toxicologists, and changing pest-control requirements in progressive food-producing economies.

These matters are of genuine concern to increasing numbers of governmental agencies and legislative bodies around the world, for some of these chemicals have resulted in a few mishaps from improper use. Adequate safety-in-use evaluations of any of these chemicals persisting into our foodstuffs are not simple matters, and they incorporate the considered judgments of many individuals highly trained in a variety of complex biological, chemical, food technological, medical, pharmacological, and toxicological disciplines.

It is hoped that "Residue Reviews" will continue to serve as an integrating factor both in focusing attention upon those many residue matters requiring further attention and in collating for variously trained readers present knowledge in specific important areas of residue and related endeavors; no other single publication attempts to serve these broad purposes. The contents of this and previous volumes of "Residue Reviews" illustrate these objectives. Since manuscripts are published in the order in which they are received in final form, it may seem that some important aspects of residue analytical chemistry, biochemistry, human and animal medicine, legislation, pharmacology, physiology, regulation, and toxicology are being neglected; to the contrary, these apparent omissions are recognized, and some pertinent manuscripts are in preparation. However, the field is so large and the interests in it are so varied that the editors and the Advisory Board earnestly solicit suggestions of topics and authors to help make this international book-series even more useful and informative.

v

"Residue Reviews" attempts to provide concise, critical reviews of timely advances, philosophy, and significant areas of accomplished or needed endeavor in the total field of residues of these chemicals in foods, in feeds, and in transformed food products. These reviews are either general or specific, but properly they may lie in the domains of analytical chemistry and its methodology, biochemistry, human and animal medicine, legislation, pharmacology, physiology, regulation, and toxicology; certain affairs in the realm of food technology concerned specifically with pesticide and other food-additive problems are also appropriate subject matter. The justification for the preparation of any review for this book-series is that it deals with some aspect of the many real problems arising from the presence of residues of "foreign" chemicals in foodstuffs. Thus, manuscripts may encompass those matters, in any country, which are involved in allowing pesticide and other plant-protecting chemicals to be used safely in producing, storing, and shipping crops. Added plant or animal pest-control chemicals or their metabolites that may persist into meat and other edible animal products (milk and milk products, eggs, etc.) are also residues and are within this scope. The so-called food additives (substances deliberately added to foods for flavor, odor, appearance, etc., as well as those inadvertently added during manufacture, packaging, distribution, storage, etc.) are also considered suitable review material.

Manuscripts are normally contributed by invitation, and may be in English, French, or German. Preliminary communication with the editors is necessary before volunteered reviews are submitted in manuscript form.

Department of Entomology   F.A.G.
University of California
Riverside, California
August 10, 1970

Table of Contents

Pesticide regulations and residue problems in Poland

By

Tadeusz Stobiecki [*]

Contents

I. Introduction

The present paper is basically concerned with pesticide residues. Production problems, assortment of pesticides, and trade legislation are, however, so closely interrelated with the problem of residues that it is deemed necessary to give short, essential information on subjects relevant to the main topic.

In the preliminary part of this review a survey is made of the situation existing in Poland in the field of pesticide residues in agricultural products. The following aspects are essential to an assessment of this situation: structure of cultures, climatic conditions, extent of application of chemicals in agriculture, assortment of pesticide chemicals used, level of agricultural practice, and, also, to some extent, social structure and organization of agriculture.

The controllable factors among these are not as yet optimizing production in this country, but they are favourable from the point of view of the level of pesticide residues forming a danger to consumers of agricultural products of both plant and animal origin.

---

[*] Plant Protection Institute, Poznań, Poland.

1

From that standpoint the existing situation in Poland is certainly better than in many West European and also Socialist countries with intense and highly mechanized agricultural practice.

The consumption of pesticide chemicals is undoubtedly the essential factor, regardless of whether it is judged by national standards or by standards of the most highly developed regional economy.

The level of this consumption is lower in Poland in comparison with many other European countries. This is best illustrated by examples of pesticide consumption/ha. of arable land in 1965: Poland ca. 0.5 kg. (concentrates), Czechoslovakia ca. 0.8 kg., German Federal Republic ca. 1.4 kg., and Great Britain ca. 1.0 kg.

Another equally important factor is formed by the types of pesticides used, proportion of toxic compounds, their stability and persistence, and methods, organization, and supervision of their application. These aspects of pesticide control will be dealt with in greater detail in further sections of this report.

II. Structure of agriculture

In 1967 the area of the land used for agriculture in Poland was 19,819 thousand ha., which included 15,518 thousand ha. of cropland (arable land and fruit plantations) and 4,301 thousand ha. of meadows and pastures.

The cropland was used in the following way, in millions of ha.: four chief wheats—8.2, potatoes—2.8, sugar beets—0.43, common (feed) beets—0.2, other feed plants—1.96, rape-seed (*Brassica napus* and *B. rapa*)—0.32, flax—0.1; the balance is field cultivated vegetables, hops, tobacco, pharmaceutical plants, etc.

The plant protection scheme covers the control of all these cultures, mainly by means of chemical treatment. The methods employed and extent of control depends on many factors. The principal factor arises from the economic structure and partitioning of farmland; large state-owned land estates constitute only 13.6 percent of the whole area, collective farms constitute 1.1 percent, and the rest consists of the individual farms.

Guiding principles for the use and extent of application of pesticides are determined every year by authorities in a central plan coordinated with the plan of plant production.

III. Information on the production and use of pesticides

The bulk of Polish industrial production is formed by pesticides for mass application. These are insecticides, fungicides, herbicides, and also certain preparations in the "miscellaneous" group. Insecticides constitute a high proportion of these chemicals, any shortage in mass materials being made up with imported products.

In the field of imports the predominant item is formed by modern selective herbicide compounds (about 47 percent of the whole plan of pesticide importation in 1969), insecticides, mainly organophosphorus (about 32 percent), organic fungicides and dressings (about 13 percent), storage disinfectants, and small quantities of certain other chemicals (about eight percent).

The long-term plans provide for some limitation of imports but not for their total elimination. This policy is vindicated by high investment costs of new production lines and poor rentability of small factories as well as specialization and cooperation within Comecon countries.

During the post-war period progress in the use of pesticides in Poland, at first slow, underwent rapid acceleration, to become nearly stabilised in recent years although, according to plans for agricultural development in this country, a further increase in the use of pesticides is to be expected in the near future.

The rate of this growth is best illustrated by the amounts of concentrates supplied in preceding years: 1955—ca. 2,000 tons, 1960—ca. 4,600 tons, 1965—ca. 8,000 tons, 1967—ca. 7,200 tons, and estimated for 1969—ca. 9,800 tons.

The essential problem, from the point of view of residues, is the nature of pesticides used and the proportion of highly toxic matter in these chemicals.

Until a few years ago insecticides formed the predominant part of the total consumption of pesticides. At present a gradual increase of less toxic herbicides and fungicides is observed.

In 1967 supplies of foreign and domestic chemicals for agricultural needs were composed of (in terms of concentrates): insecticides—51.5 percent, fungicides and dressings—18.8 percent, herbicides—25.2 percent, and "miscellaneous"—4.5 percent.

In the group of insecticides only 7.5 percent constituted I and II category poisons, more than the half of this group (four percent) were compounds for winter spraying which leave practically no residue in the crop.

According to the plan for 1969 a further decrease in the proportional content of insecticides will follow and will not exceed about 45.3 percent of total supplies of pesticides to agriculture.

The application of pesticides is, of course, closely related to the plant protection scheme.

For a better understanding of conditions existing in Polish agriculture chief pests and diseases, and also the treatment applied, are listed in the order of magnitude of treated areas:

Seed dressing—*Tilletia tritici, Ustilago avenae, U. hordei, Fusarium nivale, Ascochyta pisi, Colletotrichum lindemuthianum.*
Potato protection—*Leptinotarsa decemlineata, Phytophtora infestans.*

Chemical weed control of cultures—wheats, flax, papilionaceous
plants, beets, vegetables, orchards, nurseries.
Sugar beet protection—*Aphis fabae, Pegomyia hyoscyami.*
Protection against rodents—mainly *Microtus arvalis.*
Orchard plant protection—*Venturia inaequalis, Monilia fructigena,
Podosphaera leucotricha, Aphididae, Carpocapsa pomonella, Lasp-
syresia funebrana, Hoplocampa minuta, H. flava, Rhagoletis cerasi,
Metatetranychus ulmi.*
Rape protection—*Brassica napus, B. rapa, Meligethes aeneus, Psyl-
liodes chrysocephala, Ceuthorrhynohus* p.s.s., *Dasyneura brassicae.*
Protection of papilionaceous plants—*Laspeyrosia nigricana, Bruchus
pisorum, Apion apricans.*
Flax protection—*Longitarsus parvulus, Aphtona euphorbiae.*
Tobacco protection—*Peronospora tabacina.*
Protection of vegetables—*Phytophtora infestans, Plasmodiophora
brassicae, Peronospora destructor, Phorbia brassicae, Hylomyia an-
tiqua et cilicrura, Pieris brassicae, Phyllotrsta* sp., *Barathra brassicae.*
Hop protection—*Phorodon humuli, Tetranychus urticae, Pseudo-
peronospora humili.*

The technique of treatment may have an important influence on incidental and side effects resulting from application of chemicals in
agriculture. This applies especially to Poland, which has a highly par-
titioned rural economy. A farmer husbanding a small farm has neither
the means nor the necessary knowledge for carrying out modern and
economical plant control. Therefore in Poland the main responsibility
rests with a specialized organization fully equipped for implementa-
tion of control by a skilled staff. The final organization of these
services, involving considerable expenditure, is not yet finished. These
services include Machine Service Pools (State administered), Plant
Protection Stations, Cooperative Machine Pools, etc.

Control by specialized teams facilitates observance of the necessary
minimum interval between the last application and harvest and protection against excessive pesticide residues better than do legislating rules
and punitive measures.

The Plant Protection Institute publishes "General Recommendations on Plant Protection" once a year. This publication presents in
tabular form regulations for application of pesticides specified for various crops with lists of relevent pests, diseases, types of compounds and
their dosage, time limitations, and tolerances in force at the time of
publication. For protection of fruits and vegetables more information
may be found in publications of the Institute of Pomology and Institute of Vegetable Cultivation.

A list of pesticides which are licensed for sale is published every
year in the "Official Bulletin" of the Ministry of Agriculture.

IV. Trade legislation

According to the legislation which has been in force in Poland since January 1960 no pesticides are allowed to be sold in Poland without official approval, so that the manufacturer or the importer, according to whether the product is of domestic or of imported origin, has to obtain a certificate of registration bearing a serial number. The certificate is granted by the Ministry of Agriculture after the efficiency of the product for the intended purpose has been investigated, *i.e.* for control of the specific pest or the disease, as well as the incidental and side effects of the compound. At present a new regulation of the Ministry of Agriculture dated 22 April 1965 is in force [Monitor Polski (Polish Official Gazette) 1965 No. 28].

The main elements of this regulation are: (a) pesticides are not allowed to be sold without a licence; (b) the Ministry of Agriculture accords a licence only on the strength of a conclusive opinion from the Plant Protection Institute; (c) the Plant Protection Institute applies for approval on the basis of investigations carried out in the Institute, investigations carried out on behalf of the Institute by other specialized institutions, or the existing documentary specification; (d) the Plant Protection Institute applies for registration after having consulted with the Public Hygiene Service on problems concerning the harmful effects of the compound on the human system, and also whether the text of the label conforms with requirements of legislation on poisons; (e) registration may be permanent or temporary for a period of three years, depending upon the documentary specifications and advisability of further investigations of the registered pesticide; (f) in the event of an unfavourable opinion from the Plant Protection Institute, the Polish manufacturer has the right of appeal to a special commission; and (g) the regulation also includes principles standardizing the classification of pesticides into poison classes and establishing basic standards for packaging and selling of compounds belonging to the I class of poisons.

V. Documentation requirements
for commercial use of pesticides

The regulation described in the foregoing section includes a specification of documentation which has to be presented by the manufacturer (in triplicate) when applying for registration and licence to sell the compound for commercial use.

The documentary evidence should include: (a) results of investigations of the effectiveness of the compound as well as recommendations for application and the mode of use; (b) suggested label; (c) detailed technical description of the compound, viz., chemical analysis, instructions for use, physical and chemical properties of the compound, information concerning stability, storage conditions, types of containers or of packing material, and the mode of disposal when empty; (d) analytical methods to be applied for analysis of the compound or for determining of its residues; (e) the standard, *i.e.*, the official document confirming the elements specified in points *c* and *d*; (f) toxicological data obtained in course of investigations on acute toxicity (by mouth, by respiratory system, or by skin penetration), chronic toxicity, as well as cumulative properties of the compound (with new chemical compounds these investigations may need extension); (g) recommendations concerning the safe use of chemicals and first aid in case of poisoning; and (h) samples of the registered product.

Citizens of a foreign country applying for registration have to submit the detailed information (documentation) as specified above, through the Importing Office "Ciech" in Warsaw, except for items *b* and *e*, instead of which he must send three copies of his trade label, or instructions for use if attached to containers, and also the photostat of registration of the product in the country of origin and if possible in other countries as well. The Importing Office adds the translation into Polish of the label or the instruction of use to the application.

VI. Time limitations and tolerances

The Committee for Science and Technology, a national organization, has established special Working Commissions on the toxicology of pesticides. The Working Commissions are constituted by representatives of public institutions and research institutes concerned with the problem of pesticides. During meetings the Commission establishes the tolerances as well as minimum intervals to be observed between last application and harvest and the results are shown in tabular form. Some changes in time limitations are generally introduced into the table every year. The table of time limitations (Table I) has to be approved by the Ministry of Agriculture. The waiting periods between the latest application of a pesticide and harvest apply not only to main crops raised on the treated field, but also to intercrops and to plants on neighbouring fields if contaminated with the pesticide. The waiting periods do not apply to seeds-reproducing plantations provided that neither whole seed plants nor parts of these plants or thrashing remainders are fed to animals or used for human consumption. It is prohibited to spray or dust forage crops at any time with DDT, mixtures of DDT with BHC or TMDT, and with terpene derivative-containing formulations because these compounds accumulate in milk, milk products, meat, etc. Exceptions to this rule are indicated in comments 2, 6, and 19 of the table footnotes. If other pesticides have been applied it is allowed to use the crop for pasture or to harvest it for green forage, hay, or silage after the proper waiting period. The waiting periods do not apply also to pharmaceutical plants because that product must be completely free of pesticide residues.

The table of tolerances (Table II) is submitted for acceptance to the Minister of Health.

Decisions of the Commission are based on results of investigations carried out in Poland, technical data published abroad, and the literature.

VII. Organization of research on residues

The organization of research on pesticide residues involving hazards to man has been settled in Poland by legislation. In addition to the Regulation of the Ministry of Agriculture concerning the registration of pesticides (par. 3), the Plant Protection Act [Dziennik Ustaw PRL No 10, 23 February 1961(Official Gazette of the Polish People's Republic)], the Polish Food Act (Codex alimentarius), the Poison Act of 21 May 1963 (Dziennik Ustaw PRL No 23, 1963), and Regulations of the Ministry of Health concerning this Act. The problem of residues is a subject of a special Regulation which will come into force in 1969. According to this Regulation all Ministries concerned are committed to organize research on pesticide residues, institute special control service, and issue the necessary regulations for the enforcement of activity in this field.

Investigations on pesticide residues have been conducted in Poland for several years, however. New Regulations establish the principles of coordination of such activity and determine exactly the range and scope of responsibilities of research and control services supervised by the following ministries: Health, Agriculture, Chemistry, Food, Forestry, and Foreign Commerce. Such subdivision brings about the concentration of research activity in the hands of the most competent and qualified institutions, and under existing conditions it concerns more than ten research institutes and various university departments.

Toxicological investigations are concentrated in the Public Hygiene Service and in the Institute of Organic Industry. Public Hygiene Service also conducts investigations in the following fields: pesticide residues in food, determination of tolerances, water and soil contamination, and the application of pesticides in sanitary proceedings.

The activity of the Institute of Work Safety and Farm Hygiene is restricted to working hygiene and safety. Agricultural and technical institutes deal with problems of residues in agricultural products of plant and animal origin and with determining the dynamics of residue decay.

Agricultural institutes propose time limitations, and investigate the infiltration of pesticides used in animal hygiene, and the influence of migration and residues on biocenosis.

Table I. *Waiting periods between the latest application of a pesticide and harvest in Poland in 1969*

Active ingredient (common name)	Formulations	Waiting periods (days) (nos. in parentheses indicate specific comments [a])				
		Fruit crops	Vegetable crops	Leguminous (excluding vegetables)	Root crops	Other agricultural crops
Insecticides and acaricides						
Aldrin/TMDT	Dust	— x	— (3)	— (3)	— x	— (3)
Dieldrin/TMDT	Dust	— x	— (3)	— (3)	— x	— (3)
BHC	Liquid 10%	21 (22)	— x	21	21 (11)	21
BHC	Dust 1.2%	— x	— x	21	21 (11)	21
BHC	Dust 2.4%	— (4,9)	— (4,9)	21 (4)	21 (4,11)	21 (4)
Carbaryl	Wet. powder	7 (10,31)	7 (31)	7	7	7
Chlorfenvinfos	Liquid	— x	— x	x	21 (23)	x
Chlorfenson	Liquid	14 (10)	14	14	14	14
Chlorkamphen	Dusts, liquid	— (19)	30 (1,2,13)	30 (1,2,13,19)	30 (1,2)	30 (1,2,19)
DDT	Dust, wet. powd., liquid	— x	— (1,2,6)	— (1,2,6)	30 (1,2,6)	30 (1,2)
DDT/BHC	Dust	— x	— (1,2,6)	— (1,2,6)	30 (1,2,6)	30 (1,2)
DDT/BHC	Atomising solutions	— x	— (1,2,6)	— (1,2,6)	30 (1,2,6)	30 (1,2)
DDT/BHC	Smoke	Only in glasshouses for decorative plants				
DDT/BHC/DMDT	Dust, wet. powd., liquid	— x	— (1,2,6)	— (1,2,6)	30 (1,2,6)	30 (1,2)
Dieldrin	Wet. powder	— x	— (18)	— x	— x	x
DMDT/methoxychlor	Liquid	7 (10)	14	7	7	7
Dichlorfos	Atomising solutions [b]	— x	— x	— x	7	7
Dichlorfos	Liquid	7 (10)	7 (27)	7	7	7
Dimethoate	Liquid	21 (14)	21 (26)	21	21	21
Disulfoton	Granular	— x	— (5)	— x	— (5)	— x
Demeton-methyl	Liquid	42 (1,14)	42 (1)	42 (1)	42 (1)	42 (1)
Demeton-S-methyl	Liquid	30 (1,14)	30 (1)	30 (1)	30 (1)	30 (1)
Endosulfan	Liquid	30 (10,28)	30 (28)	30 (28)	30	30 (28)

Fenitrothion	Liquid	21 (10,15)	14	14	14	14
Fenitrothion	Dust	21 (10)	14	14	14	14
Fention	Liquid	42 (1,30)	x	x	x	x
Formothion	Liquid	21 (15)	21	x	x	21
Malathion	Atomising solutions [b]	x	5 (7)	7	7	7
Malathion	Liquid	7 (7,10)	7 (7)	21 (1)	21 (1)	21 (1)
Parathion-methyl	Dust	x	x	21 (1)	21 (1)	21 (1)
Parathion-methyl	Liquid, wet. powder	21 (1,12)	x	x	7 (23)	x
Proxopur	Liquid, wet. powder	x	x	14	14	14
Tetradifon	Liquid	14 (10)	14	60 (1)	60 (1,25)	60 (1,25)
Tiometon	Liquid, wet. powder	x	x	14	14	14
Trichlorfon	Liquid	14 (10)	10	14	14	14
Fungicides						
Copper	Wet. powders	7	7 (20)	7	7	7
Captan	Wet. powder	7 (8)	7	7	7	7
Chinomethionat	Wet. powder	21	21 (20)	x	x	x
Dodine	Wet. powder	14	x	x	x	x
Dinocap	Liquid, wet. powder	21	21 (20)	x	x	21
Dichlofluanid	Wet. powder	14 (21)	14	x	x	14
Fentin acetate [c]	Wet. powders	x	x	x	42 (17)	x
Maneb	Wet. powder	7 (8,16)	7	7	7	7
Metiram	Wet. powder	7	3	7	7	7
Tiocyan-dinitrobenzene	Wet. powder	21 (8)	x	x	21	21
TMDT (Thiram)	Wet. powder	7 (8)	7	7	7	7
Zineb	Wet. powder	7 (8,16)	7	7	7	7
Herbicides						
2-4-D, MCPA, Barban	Liquid, powders	x	(24)	(24)	(24)	(24)
Dinitrobutyl-phenyl-acetate	Wet. powder	x	42	42	x	42
Molluscicides						
Acetaldehyde (Meta)	Liquid, powder	14	14	14	14	14

Table I. (continued)

ᵃ Specific comments and explanations to numbers in parentheses:

x = the pesticide is not used in this culture.

1 = It is not recommended for use in home gardens and garden plots.

2 = it is not allowed to apply on pastures and forage crops; it can be used in sugar beets in early growth stages (no more than 10 leaves) and on potato plantations; waiting period for treated beet greens, if used as forage, amounts to 30 days after dusting and to 60 days after spraying with emulsion or suspension; treated potato greens must not be fed to animals.

3 = only for seed dressing, except carrot, radish, turnip, and other root crops.

4 = if used for soil treatment (raising of carrots, potatoes, and other root crops) is not allowed until the fourth year after application.

5 = recommended only for soil treatment, particularly on seed-producing plantations; on sugar beet fields can be applied at sowing time or later until the stage 4–6 leaves/seedlings, in that case beet greens cannot be fed to animals until 60 days after application; in hop plantations it can be applied only at the first hoeing.

6 = in vegetable and turnip and cabbage (rutabagae) plantations can be applied only within 10 days after germination or seedlings planting; in leguminous crops only within 14 days after germination.

7 = not to be applied on plum and tomato fruits because it may produce fruit off-flavour.

8 = not to be applied on strawberry fruits because it may produce fruit off-flavour.

9 = only for soil pest control in fruit tree nurseries and on small fruit plantations; in vegetable production for seedling dipping.

10 = on strawberries and raspberries only before blossoming and/or after harvest.

11 = not to be applied on potato plantations because it may produce an off-flavour.

12 = 21-day waiting period only for plums and prunes, other fruit trees can be sprayed before blossoming or shortly after petal fall; small fruits can be sprayed only before blossoming and/or after harvest.

13 = only in broad bean plantations (for forage and human consumption) during blossoming.

14 = on apples, pears, and plums only within 14 days after petal fall; on sweet cherries immediately after petal fall (except early varieties which can be sprayed only after harvest); on currants, gooseberries, strawberries, and raspberries only after harvest.

15 = if used on sweet cherries against cherry fruit fly, waiting period can be decreased to 14 days.

16 = on currants only within 14 days after blossoming.

17 = on potato no waiting period is required but greens are not to be used as forage.

18 = only for onion seed dressing.

19 = if applied against rodents, waiting period is 42 days; on forage crops only in autumn, not later than 150 days before harvest.

20 = 4-day waiting period on cucumbers.

21 = 42-day waiting period on wine grapes.

22 = on apples, pears, and plums can be used before blossoming or immediately after petal fall; on cherries (sweet and sour) and on small fruits only before blossoming or/and after harvest.

23 = only on potato plantations, greens are not to be used as forage.

24 = the sprayed plants should not be fed to animals before 21-day waiting period.

25 = 30-day waiting period for sugar beets, mangel, and hops.

26 = if used for soil application by watering, waiting period for cabbage and onion is 60 days, for kohlrabi, 42 days; radish can be watered only at germination time.

27 = not to be applied in glasshouses where lettuce is raised; in mushroom houses not later than 5 days before harvest.

28 = for rodent control, waiting period is 60 days.

29 = application for watering of radish allowed only at germination time.

30 = only a single application on apple trees.

31 = 14-day waiting period for plums and green vegetables; not to be applied on cabbage and other cruciferous vegetables.

b In glasshouses.

c Or hydroxide.

Table II. *Tolerances of residues in fruits and vegetables in Poland accepted for temporary use in 1969*

Active ingredient (common name)	Tolerance (p.p.m.)	Active ingredient (common name)	Tolerances (p.p.m.)
Aldrin	0.01 [b]	Captan	15
Chinomethionat [a]	0.3	Dinitrotiocyanbenzene	2
Chlorfenson	1.5	Fenitrothion	0.4
Chlorkamphen [a]	0.4	Formothion [a]	0.5
Carbaryl	3	Lindan (BHC)	2
Copper (as Cu)	10	Malathion	3
DDT	1	Maneb	3
Demeton-methyl		Methoxychlor	10
and S-methyl	0.4	Parathion-methyl	0.5
Dichlofluanid [a]	0.4	Sulphur	50
Dichlorfos	0.2	Tiometon [a]	0.5
Dieldrin	0.01 [b]	TMDT (Thiram)	3
Dimethoate	0.5	Tetradifon	1
Dinocap [a]	1	Trichlorfon	0.5
DNCK	—	Zineb [a]	3
Dodine [a]	1		

[a] Proposed by the commission.
[b] Zero for technical grade material.

VIII. Organization of residue control and enforcement of time limitations

The control of pesticide residues is carried out in several ways depending upon the controlled object.

A number of chemical laboratories belonging to Regional Sanitary and Epidemic Stations carry out random control of alimentary products sold within this country. They also make random control of water, effluents, and soil.

Random control of pesticide residues in plant and animal products at the place of production by laboratories of the Ministry of Agriculture is in course of organization. Some of these laboratories have been in existence for some time, others, in number sufficient to provide for all the needs of the country, will be established in the near future.

The enforcement of time limitations is supervised throughout the country by dozens of District Plant Protection Stations administered by local authorities.

The control of materials processed in the alimentary food industry is made in laboratories belonging to that industry. The control of export foods (agricultural products or processed alimentary products) is carried out in laboratories of the Ministry of Foreign Commerce under conditions settled by the terms of the contract.

IX. Conclusions

The problem of pesticide residues in Poland presents so far no great hazards, owing to a moderate application of chemicals in agriculture.

A great importance is nevertheless attached to the problem of residues in this country on account of safety and health of the population, of plans for further intensification of agricultural production, and also in view of increasing stringency in demands of the export trade.

The problems connected with pesticide residues and their control in various branches of the national economy are an object of coordinated investigations and organization measures based on legislative regulations and official recommendations.

Problems of toxicology and residues require a concerted collaborative effort of international organizations for settling various approaches. Poland is actively cooperating with some international organizations concerned with plant protection and *inter alia* with FAO, WHO, EPPO, and also with respective branches of COMECON countries. Polish research institutions are willing to take an active part in this collaborative effort and to exchange information on the results achieved in the course of research and of practical experience.

Summary *

The structure of agriculture in Poland is described to provide background for the production of, use of, and restrictions on pesticides and their application in Polish agriculture. Requirements for licensing and registration are discussed, and timing restrictions (minimum intervals) and existing tolerances are presented in detail.

The organization of Polish residue research and of tolerance and time limitation enforcement is described.

Résumé **

La Législation des pesticides et les problèmes de résidus en Pologne

La structure de l'agriculture polonaise est décrite pour expliquer la production des pesticides, leur emploi et les mesures restrictives à leur utilisation en Pologne.

Les exigences pour l'autorisation et l'enregistrement des formulations sont présentées; les tolérances existantes ainsi que les intervalles d'interdiction avant les récoltes sont donnés en détail.

Le mémoire décrit l'organisation de la recherche sur les résidus en Pologne, ainsi que les méthodes de surveillance du respect des tolérances et des délais d'utilisation.

---

* Prepared by the Editor.
** Traduit par R. MESTRES.

Zusammenfassung *

Gesetzliche Handhabung von Pestiziden und Rückstandsprobleme in Polen

Die Struktur der Landwirtschaft in Polen wird beschrieben sowie, vor diesem Hintergrund, Produktion, Einsatz und behördliche Kontrolle der Pestizide und deren Anwendung in der polnischen Landwirtschaft. Die Anforderungen für die Anerkennung und Registrierung werden diskutiert; zeitliche Regulierungen (Wartezeiten) und gültige Toleranzen werden eingehend dargestellt.

Die Organisation der Rückstandsforschung und der Kontrolle über die Einhaltung von Toleranzen und Wartezeiten in Polen wird beschrieben.

---

* Übersetzt von H. FREHSE.

Interaction of pesticides with
aquatic microorganisms and plankton

By

GEORGE W. WARE * and CLIFFORD C. ROAN *

Contents

I. Introduction

The interactions of pesticides [1] and soil microorganisms are heavily documented from the agricultural view. Many of the same soil particles, microorganisms, and pesticides are found in freshwater and estuarine ecosystems, and similar relationships may exist. It is the purpose of this paper, then, to review the interactions of pesticides and aquatic microorganisms, those microscopic plants and animals found in freshwater, estuarine, and marine environments.

a) Definition

Aquatic microorganisms and plankton, by the authors' definition, are comprised of those microflora and microfauna commonly found in

---

* Department of Entomology, University of Arizona, Tucson, Arizona. Journal Series #1558.

[1] Pesticide chemicals mentioned in text are identified chemically in Table XVI.

fresh water, brackish, and marine environments. These include (1) the higher protists, or unicellular organisms, which are the algae, protozoa, fungi, and slime molds, and (2) the lower protists, blue-green algae, myxo-bacteria, spirochetes, and eubacteria. Plankton are organisms found floating or drifting almost passively and are carried about by wave action and currents. The phytoplankton are composed generally of algae to which the diatoms belong. Zooplankton are mostly very small and belong generally to the Protozoa, and animal phyla Arthropoda, Porifera (sponges), Cnidaria (coelenterates), Platyhelminthes (flatworms), Aschelminthes (roundworms), and Annelida (segmented worms). Plankton may also involve immature or larval stages of organisms normally excluded from this category, as well as microflora and fauna relocated from their original terrestrial surroundings.

Aquatic microorganisms are in many instances "weeds" in that they are merely soil microorganisms out of place. Fresh surface waters are usually not identified with any characteristic bacterial flora (FROBISHER 1949). The kinds of bacteria in them depend on their mineral and organic content, the soils with which they are in contact, surface pollution, and other factors. Many of the higher bacteria are common in fresh water lakes. Rainfall increases the numbers of soil bacteria in waters which collect runoff. Waters polluted with sewage have broad population ranges of *Escherichia coli* and other Enterobacteriaceae as well as Enterococci and *Clostridium perfringens.* Fecal pollution contributes such soil saprophytes as *Spirillium, Sarcina, Micrococcus, Mycobacterium, Bacillus,* yeasts, molds, Actinomycetes, and many others. The extent and nature of the bacterial populations in streams and rivers where sewage treatment plant effluents are found should differ significantly from mountain lakes and streams derived from melting snow.

b) Entry of pesticides into aquatic environments

WESTLAKE and GUNTHER (1966) classified environmental contamination under two categories, *i.e.*, (1) intentional (direct) application and (2) unintentional (indirect) contamination. In Table I (WESTLAKE and GUNTHER 1966), we have reclassified these sources for the aquatic environment. Our classification is perhaps more stringent (or the result of better hindsight) than the above-referenced authors in that many of the sources they classed as unintentional we have classified as intentional. This classification, we feel, is justified because pesticides were in fact knowingly added to the aquatic environment. What was unknown was their potential persistence and the as yet incompletely comprehended long range consequences of such additions.

For instance, the association of run-off pesticides with organic particulate matter in estuaries is significant because many organisms rely on these particles for part or all of their energy requirements. Accord-

Table I. *Sources of pesticides in the aquatic environment*

A. Intentional introductions
   1. Control of objectional flora and/or fauna
   2. Industrial wastes
      a. Pesticide manufacturers and formulators
      b. Food industry
      c. Moth-proofing industry
   3. Disposal of unused materials
   4. On-site field cleaning of application, mixing, and dipping equipment
   5. Disposal of commodities with excessive residues
   6. Decontamination procedures
B. Unintentional introductions
   1. Drift from pesticide applications to control objectional flora or fauna
   2. Secondary relocation from target area via natural wind and water erosion
   3. Irrigation soil water from target areas
   4. Accidents involving water-borne cargo
   5. Application accidents involving missed targets or improper chemicals

ing to ODUM *et al* (1969) plant detritus consumers include amphipods, isopods, harpacticcid copepods, various filter- and deposit-feeding bivalves, annelid worms, caridean and penaeid shrimp, fiddler crabs, and mullets. DDT and its metabolites accumulate in plant detritus within estuaries and may persist there for many years. The residues appear to be most abundantly associated with particulates having diameters from 250 to 1,000 microns (μ).

A detailed recitation of the pesticides detected in the aquatic environment at various places and times would include all pesticides that have been or are in current use. The persistence of various pesticides in the nonliving components of the aquatic environment, while beyond the scope of the present review, must be recognized particularly where this persistence may be greater than in the soil. A specific example, from SCHWARTZ (1967), of the metabolism of 2,4-D being much slower in aqueous than in soil environments is illustrative of this situation.

In addition to possible physical/chemical reactions of pesticides in water with the nonliving components two other interaction systems are of interest, *i.e.*, the direct action of the pesticide on the flora and fauna of the environment, and the effects of the flora and fauna on the pesticides.

II. Toxicity of pesticides to aquatic microorganisms

Toxicity is frequently regarded only as the lethal effects of the chemical upon a particular organism. Other direct effects that are included in this review concern changes in growth rate and changes in specific metabolic rates, *i.e.*, photosynthesis. Indirect actions that are ecologically important are the results of stress on one or more organisms that permit previously suppressed competitors to flourish, further stressing dominant populations.

The diversity of species in both the fresh water and marine environments is even greater than the diversity of chemicals added to these environments. This diversity of organisms has several important consequences giving rise first of all to a necessary multiplicity of experimental systems and techniques for evaluation of the toxicity of environmental contaminants. A further consequence of major importance is the tendency, particularly of the public (and only slightly less on the part of the scientist), to extrapolate from the adverse effects of certain chemicals or classes of chemicals on a few organisms to include an entire ecosystem.

The diversity of laboratory tests is represented by the designs employed by Wurster (1968) where cultures of the test phytoplankton were exposed to different concentrations of DDT for 20 to 24 hours, with 14 hours of light and ten hours of dark. After this exposure period C^{14}-bicarbonate was added and the algae were illuminated for an additional four to five hours. Radio-assay of the filtered cultures was employed to estimate the amount of carbon fixed by photosynthesis. The effects of DDT exposures were determined by comparisons of the exposed and control populations.

In contrast, Ukeles (1962) prepared the toxicants as concentrated or saturated stock solutions in sterile sea water. After at least 24 hours, serial dilutions in sterile sea water were prepared and appropriate amounts were added to the sterile basal medium. Cultures were incubated for ten days under continuous illumination. Growth rates were estimated by determination of transmittance at 530 mμ.

The excellent tabulation of Lawrence (1962) in the data pertaining to microflora and fauna and planktonic forms in general lists such diverse test environments as aquaria, distilled water, mud, and ponds, with exposures to the herbicides ranging from a few hours to 60 days. Effects are described as lethal, percent killed, not toxic, and tolerated.

The effects of a series of pesticides on estuarine phytoplankton were investigated by Butler (1965 a and b). Some of these results, together with data from Circular 167 (1963) appear in Table II. These data were based on a four-hour exposure to the toxicants in question. The organisms were described only as phytoplankton. Pesticides at concentrations of 1.0 part per million (p.p.m.) reducing carbon fixation by 75 percent or more are comprised of six herbicides, 11 insecticides (two organophosphorus and nine organochlorine), and two fungicides. Of the 26 herbicides in Table II, nine are listed as having no adverse effect under the test conditions. Of 25 insecticides and the three fungicides tested all had adverse effects on phytoplankton.

Further data on the effects of DDT on photosynthesis by four species of phytoplankton are presented in Table III (Wurster 1968). These data suggest that a specific adverse effect from DDT concentrations as low as 0.1 p.p.m. may be expected. Wurster (1968) suggests the absence of a threshold concentration of DDT on the basis of a

Table II. *Effects of pesticides at concentrations of one p.p.m. on estuarine phytoplankton*

Pesticides [a]	Decrease of carbon fixation during 4 hours exposure (%)
Herbicides [b]	
Monuron (urea)	94
Neburon (urea)	90
2,4,5-T, polyglycol butyl ether ester (phenoxy)	89
Diuron (urea)	87
Silvex (phenoxy)	78
DEF (defoliant)	75
Zytron (phosphate)	59
Paraquat	53
2,4-D, 2-ethylhexyl ester	49
Diquat	45
2,4-D, propyleneglycol butyl ether ester	44
Fenuron	41
Dacthal	37
Tillam	24
2,4-D, butoxy ethanol ester	16
N-Serve	15
Hydram	9
Dalapon Na salt	0
Kurosal SL (60 percent Silvex)	0
Tordon	0
2,4-D acid	0
2,4-D dimethylamine salt	0
2,4,5-T acid	0
MCP amine	0
Eptam	0
Vernam	0
Insecticides	
Kepone	95
Heptachlor	94
Chlordane	94
Toxaphene	91
Ronnel	89
Thiodan	87
Methyl Trithion	86
Dieldrin	85
Aldrin	85
Methoxychlor	81
DDT	77
Ethion	69
Dibrom	56
Di-Syston	55
Endrin	46
Mirex	42
Bayer 37344	39
ASP-51	30
Lindane	28
Carbaryl	17

Table II. (continued)

Pesticides [a]	Decrease of carbon fixation during 4 hours exposure (%)
Imidan	8
Demeton	7
Baytex	7
Malathion	7
Diazinon	7
Fungicides	
Ferbam	97
Dyrene	91
Phaltan	32

[a] Insecticides and some other materials from ANONYMOUS (1963).
[b] Herbicides from BUTLER (1965 a).

sigmoid dose-response curve in the case of *Skeletonema costatum* where the concentration of DDT was constant but the cell population density varied. This suggestion does not appear warranted in view of the presented data indicating the absence of any clearly definable effect at concentrations at or below 1.0 part per billion (p.p.b.).

UKELES (1962) presented data on the toxicity of 17 chemicals to five species of marine phytoplankton. Table IV is a rearrangement of data from this paper to indicate the highest concentrations preventing any growth of the cultures during a period of ten days. These data indicate that of the insecticides tested only toxaphene approaches the phytotoxicity of herbicides tested against marine phytoplankton.

PIERCE (1958 and 1960) found that Kuron applied to a fresh water pond had no apparent lasting effect on various plankton although there was a noticeable decrease in populations within 24 hours after treatment. The predominant forms observed were *Ceratium, Dinobryon, Synura,* and *Bosmina,* while *Spirogyra, Volvox, Daphnia, Tabellaria, Micrasterias,* and *Fragilaria* were frequently taken.

Table III. *Effects of DDT on photosynthesis by phytoplankton*

Organism	C^{14} uptake as % of controls DDT concentration	
	1.0 p.p.b.	100 p.p.b.
Skeletonema costatum	70	25
Coccolithus huxleyi	75	35
Pyramimonas sp.	80	20
Peridinium trochoideum	55	25

Table IV. *Concentrations of certain pesticides preventing growth in five species of marine phytoplankton*

Pesticide	Concentration in p.p.m. producing no growth				
	Proloccus sp.	*Chlorella* sp.	*Dunaliella euchlora*	*Phacodac- tylum tricornutum*	*Monochry- sis lutheri*
Trichlorfon	1000	500	500	500	100
TEPP	500	500	500	500	500
Phenol	500	500	500	100	100
Dowacide A	100	100	100	100	50
Orthodichlorobenzene	13	13	13	13	13
Chloronitropane	80	80	80	80	80
PVP-iodine	>100	>100	50	50	100
Carbaryl	10	10	10	< 0.1	1
Nabam	10	10	1	1	1
Lindane	> 9	> 9	> 9	7.5	7.5
Toxaphene	.15	.07	0.15	.04	.01
DDT	> 60	> 60	> 60	> 60	60
Lignasan	.006	.006	.06	.06	.006
Fenuron	29	2.9	2.9	2.9	2.9
Neburon	0.2	0.2	0.2	0.2	<0.04
Monuron	0.02	0.02	0.02	0.02	0.02
Diuron	0.004	0.04	0.004	0.004	$<2 \times 10^{-5}$

COWELL (1965) in pond studies in New York observed that neither Silvex nor sodium arsenite was toxic or inhibitory to zooplankton at concentrations of two p.p.m. Zooplankton populations were drastically reduced by applications of nine p.p.m. of sodium arsenite. COPE (1966), however, in studies of a fresh-water ecosystem found sodium arsenite to have an LC_{50} of 1,800 p.p.m. for *Daphnia pulex*.

Selected data from COPE's (1966) investigations are presented in Table V. As might be expected the insecticides are much more toxic to *Daphnia* than are the herbicides. HARDY (1966) indicates that the herbicide Tordon at one p.p.m. was nontoxic to *Daphnia*.

TATUM and BLACKBURN (1962) indicate that diquat at 0.5 p.p.m. adversely affects plankton but that recovery is rapid. MULLIGAN (1967) lists copper sulfate, potassium permanganate, combinations of copper sulphate and silver nitrate, and simazine as nontoxic to zooplankton at concentrations that are toxic to phytoplankton. CABEJSZEK and STANIS-LAWSKA (1967) report Metasystox tested on organisms found in the Vistula River to be most toxic to *Daphnia* and that algae are the least sensitive. In investigations with methyl parathion in the same environment CABEJSZEK and STANISLAWSKA (1966) also found protozoa and daphnids to be more susceptible than filamentous algae.

Further evidence of the apparent toxicity of insecticides to aquatic plants as well as animals is found in LAZAROFF (1966) where algal development is inhibited by one p.p.m. of DDT or lindane as well as

Table V. *Relative toxicities of pesticides to Daphnia pulex*

Pesticide	Toxicity (48 hr. LC$_{50}$ in p.p.m.)
DDT	0.4
Diazinon	0.9
Malathion	2
Azinphosethyl	3
Carbaryl	6
Toxaphene	15
Endrin	20
Pyrethrins	25
Heptachlor	42
Dieldrin	250
Lindane	460
Trifluran	240
Diuron	1,400
Sodium arsenite	1,800
Silvex	2,400
2,4-D	3,200
Dichlobenil	3,700
Fenac	4,500

the fungicides captan or Nabam. He found that thiocarbamate pesticides interfere with photo-induced development in the blue-green algae *Nostoc muscorum* A. Although parathion at concentrations of ten p.p.b. inhibits motility in *Euglena gracilis,* it does not prevent growth.

Investigations of the effects of pesticides on bacteria in aquatic environments are not so numerous. LICHTENSTEIN *et al.* (1966) include data on the abundance of bacteria in water stored for three months containing various insecticides. Table VI (LICHTENSTEIN 1966) indicates that DDT apparently does not have an adverse action, as do the

Table VI. *Changes in bacterial populations after storing for three months in one p.p.m. aqueous solutions of pesticides* (LICHTENSTEIN 1966)

Pesticide	Bacteria/ml. in control bacteria/ml. in treatment	
	Lake water	Soil water [a]
Parathion	0.43	0.62
Methyl parathion	1.14	0.62
Azinphosmethyl	0.25	0.47
Dieldrin	2.46	0.41
DDT	1.57	2.04
BHC	0.17	0.57

[a] Water percolated through soil before adding pesticides.

other compounds listed, on bacteria derived from soil, while in the case of lake water only parathion, Guthion, and BHC have adverse effects. Luczak and Maleszewska (1967) report that Metasystox at seven p.p.m. does not significantly affect the growth rate of bacteria.

Butler and Springer (1963) cite experiments with chronic exposures of three to six months (Table VII) in which test animals were maintained in running sea water to which the pesticide was continuously added. These authors state, "The unfiltered sea water supplying these aquaria contained planktonic larvae. Large numbers of these benthic animals, including at least 25 species in 7 phyla, set fortuitously and grew equally well in the experimental and control aquaria." Jones and Moyle (1963) found that average counts of cladocerans, copepods, ostracods, rotifers, and *Volvox* were of the same general magnitude for treated and control ponds following field tests with DDT at one lb./acre for mosqutio control.

Kasahara (1962) found that Dipterex at ten p.p.m. had no adverse effects on phytoplankton. Concentrations of 0.2 p.p.m. killed all crustacea except certain copepods while rotifers, flagellates, and ciliates were not affected.

Copper is an essential minor element for microorganisms and is commonly added in concentrations of 0.2 to 2.0 p.p.m. to synthetic culture media, and as a general rule, bacterial numbers increase with organic matter in sediments as well as in soils. However, residual copper from copper sulfate weed treatment in two Oregon coastal lakes effectively decreased the numbers of bacteria in bottom samples taken two years later, according to Watson and Bollen (1952). This inhibition was probably caused by 11 to 30 p.p.m. of total copper and occurred in the presence of large amounts of organic material. Controls (four to seven p.p.m. of copper) gave much higher counts even when organic matter was low. The results are anomolous considering the generally low toxicity of copper for bacteria, plus the copper-binding characteristic of organic matter.

Crance (1963) stated that the reduction of blue-green algae in a

Table VII. *Concentrations of insecticides in running sea water*

Pesticide	Concentration (p.p.m.)	Test animal
DDT	1	Juvenile clam
Dieldrin	5	Juvenile clam
Dieldrin	0.1	Spot
Dieldrin	0.01	Spot
Dieldrin	0.001	Spot
Aldrin	2	Oysters and mussels
Malathion	2	Oysters and mussels
Toxaphene	50	Oysters and mussels

fresh water pond should be beneficial to fish, in that water bacteria usually increase enormously after algae decay. Protozoa then feed on the bacteria followed by an increase in Rotifera and Crustacea used by fish as food. The number of zooplankters usually increased within a few days after an application of copper sulfate (CRANCE 1963). Copepods were the predominant species, followed by cladocernas. Rotifers, Chaoboridae larvae, and ostrocods were the remaining zooplankters, and were usually more abundant after applications of copper sulfate.

Toxicity, when expressed inversely, is the flourishing of one group of microorganisms with the suppression of others. A distinct increase in the blue-green algae populations was observed when lindane was applied at five, six, and 50 kg./ha. to submerged Philippine rice soils (RAGHU and MACRAE 1967 a), which they attributed to elimination of small algae-eating animals. They also observed (1967 b) that lindane at six kg./ha. resulted in significant increases in nitrogen fixation, and higher populations of anaerobic, phosphate-dissolving bacteria.

Diazinon applied at two or 20 kg./ha. was found to significantly stimulate the actinomycete and the algal populations in submerged Philippine rice soils (SETHUNATHAN and MACRAE 1969). Since most investigations seek and anticipate population decreases there may be many unobserved flourishing species in such tests.

III. Concentration

a) Freshwater microorganisms

The water solubilities of most pesticides, particularly the organic insecticides, are generally quite low varying from about one p.p.b. to complete miscibility with water. Their relative solubilities from least to greatest are the organochlorine, carbamate, and organophosphate insecticides (GUNTHER et al. 1968).

Pesticides, partly because of their very low water solubilities, tend to seek living organisms. Microscopic plants and higher aquatic florae quickly accumulate quantities of pesticides from the water medium and retain them in and on their tissues. Aquatic fauna of all sizes similarly tend to remove pesticides from the water and store them (COPE 1966).

In certain macroflorae submersed plant leaves accumulated relatively high amounts of the herbicides endothall or diquat from 0.1 p.p.m. concentrations. Disodium endothall was absorbed more rapidly and in larger amounts by leaves of susceptible plants, American and sago pondweed (*Potamogeton nodosus* and *P. pectinatus*) than by leaves of the resistant plant, American elodea (*Elodea canadensis*). Elodea leaves, however, absorbed more diquat than the sago (SEAMAN and THOMAS 1966).

In still smaller plants which begin to approach the definition of "aquatic microorganisms," WARE *et al.* (1968) observed that a filamentous alga and a pondweed (Cladophora and Potamogeton) were better indicators of surface irrigation water contamination than the alga Ocillatoria.

Bottom fauna in the deep water of the Great Lakes are dominated by the amphipod, *Pontoporeia affinis*. This species is an important food source for fish and long-tailed ducks in Lake Michigan. Pooled samples of this crustacean averaged 0.41 p.p.m. for DDT, DDE, and TDE, or 50 times the level found in surrounding muds. Fish which fed chiefly on the amphipods, namely alewives, chub, and whitefish, had about ten times, while the ducks had 15 times that of *Pontoporeia* (HICKEY *et al.* 1966).

Three species each of fungi, streptomycetes, and bacteria were tested for their ability to accumulate DDT and dieldrin from distilled water (CHACKO and LOCKWOOD 1967). After four hours the fungi accumulated 75 percent of the dieldrin and 60 to 83 percent of the DDT. One bacteria, *Agrobacterium tumefaciens,* accumulated 90 percent of the dieldrin and 100 percent of the DDT, while the results from the others were inconclusive, probably due to losses through cell washing. Using an autoclaved *Streptomycetes,* the authors concluded that heat-killed and living mycelia accumulated most of the insecticides from the medium, indicating that the accumulation was probably physical and did not involve metabolism. The accumulation of pesticides by microbes in soil and subsequently in water, is probably a strong factor in the retention of these compounds.

Cultures of a blue-green alga (*Anacyctis nidulans*), a green alga (*Scenedesmus obliquus*), a flagellate (*Euglena gracilis*), and two ciliates (*Paramecium bursaria* and *P. multimicronucleatum*) were exposed separately to DDT or to parathion at one p.p.m. (GREGORY *et al.* 1969). These algae and protozoa concentrated DDT 99 to 964 times and parathion 50 to 116 times during a seven-day exposure period. *P. multimicronucleatum* absorbed the highest levels of both insecticides. This indicated that an organism feeding on these unicellular forms may receive a higher level of insecticide than directly from water. Only small amounts of DDT or parathion remained in the supernatent liquid after seven days. No metabolities were detected of either compound in any of the organisms. The possibility of insecticide metabolism having occurred should not be excluded considering the one-step hexane extraction method employed.

In a similar study, using mycelial fragments and soil instead of a water medium, the accumulation and concentration of dieldrin, DDT, and the herbicide PCNB above ambient levels of mycelia of actinomycetes and fungi was demonstrated (Ko and LOCKWOOD 1968). However, in soil the total amount of these compounds accumulated by mycelia was relatively small, the highest being ten percent when a large amount of mycelium was used. This indicates that the ability to take up chlorinated pesticides may be a generally nonspecific property of cells of actinomycetes, bacteria, and fungi. Here the differences in the accumulation or absorption from their respective media are explained by the competitive adsorption by soil particles and a lack of continuing surface exposure to soil as in the aqueous medium.

Copper sulfate was absorbed within 72 hours and removed from solution by a heavy bloom of algae. These ponds in New Jersey were treated to yield 0.5 p.p.m. of copper but it will be noted in Table VIII

Table VIII. *Typical concentration of copper in New Jersey ponds treated with copper sulfate to yield 0.5 p.p.m. of copper* (Toth and Riemer 1968)

Water depth	Concentration (p.p.m.) after					
	Initial	2 hr.	4 hr.	24 hr.	48 hr.	72 hr.
Top	0.05	0.14	0.14	0.13	0.11	0.11
Bottom	0.03	0.06	0.06	0.13	0.13	0.09

(Toth and Riemer 1968) that at no time did the actual concentration approach the intended level.

b) Marine microorganisms

Pesticides applied directly to estuarine waters are absorbed by plankton almost immediately (Butler 1967). Here it was observed that the biological magnification of residues in the food web could progress from an estimated 1.0 p.p.b. of DDT and related metabolites in the water to 70 p.p.b. in plankton to 15 p.p.m. in fish and up to 800 p.p.m. in porpoise blubber.

A marine diatome (*Cylindrotheca closterium*) absorbed and concentrated DDT up to 190-fold from its medium containing 0.1 p.p.m. of DDT (Keil and Priester 1969).

DDT residues in the soil of an extensive Long Island salt marsh averaged more than 13 lb./acre, with a maximum of 32 lb. A systematic sampling of various organisms showed concentrations of DDT increasing with trophic level through more than three orders of magnitude. When the bottom mud held 0.28 p.p.m., zooplankton contained 0.04 p.p.m. and the ringbilled gull 75 p.p.m. (Woodwell et al. 1967).

Measuring dieldrin and DDE in micro- and macrozooplankton off the Northumberland Coast, Robinson et al. (1967) concluded that there appeared to be a close correlation with residues and their trophic level. The lowest residues were found in the second while the highest were recovered from animals in the fifth. From their survey they determined that marine animals did not appear to be better indicators of environmental contamination by insecticides than terrestrial animals.

IV. Metabolism of Pesticides

Because of the previously established relationship of many soil and aquatic microorganisms, the discussion must again be somewhat all-inclusive. That is, if pesticide metabolism occurs with a particular organism in its terrestrial environment quite naturally it should occur similarly qualitatively, though not necessarily quantitatively, in its aquatic surroundings. The essential differences appear to be usually fewer nutrients/unit mass in water than soil, thus probably less biochemical activity, and the much more rapid food chain turnover in the aquatic than in the terrestrial environment.

Many pesticides are resistant to microbial action, and either they remain unaltered, even in the presence of a large and active microbial population, or they are metabolized at a disturbingly slow rate (Table IX) (ALEXANDER 1964).

Table IX. *Persistence of insecticides in some soils* (ALEXANDER 1964)

3 years	5 years	11 years
DDT	Parathion	BHC
Dieldrin	Lead arsenate	Chlordane
Toxaphene		

Several good reviews have been compiled relating to the interaction of soil microorganisms and pesticides (MARTIN 1966, ALEXANDER 1964, EDWARDS 1966). Microorganisms in soils decompose or metabolize many of the herbicides (AUDUS 1964, FUNDERBURK and BOZARTH 1967, THEIGS 1962, MACRAE and ALEXANDER 1965), fungicides (MUNNECKE 1966), organophosphate insecticides (AHMED and CASIDA 1958, GUNNER and ZUKERMAN 1968, ROBERTS et al. 1962), organochlorine insecticides (CHACKO et al. 1966 and 1967, EDWARDS 1966), and antibiotics (GOTTLIEB 1952, PRAMER 1958).

Many species of protists have been identified which metabolize herbicides and insecticides (AUDUS 1964, MUNNECKE 1966, THIEGS 1962, ALEXANDER 1964). A list of genera represented in these reviews is shown in Table X.

a) Organochlorine insecticides

Yeasts are universal in distribution and are known to metabolize DDT (KALLMAN and ANDREWS 1963). They observed that C^{14}-DDT was converted only to DDD (TDE) by reductive dechlorination. No DDE was observed, and when DDE was incubated with yeast cultures only DDE was recovered.

The marine diatome, *Cylindrotheca closterium*, converted DDT only to DDE (nine percent) after three weeks in a medium containing

Table X. *Genera of selected microorganisms known to metabolize pesticides*

Bacteria	Actinomycetes	Fungi	Algae	Yeasts
Achromobacter	Mycoplana	Acrostalagmus	Chlamydomonas	Saccharo-
Aerobacter	Nocardia	Aspergillus	Chlorella	myces
Agrobacterium	Streptomyces	Clonostachys	Cladophora	Torulopsis
Alcaligenes		Cylindrocarpon	Cylindrotheca	
Arthrobacter		Fusarium	Oscillatoria	
Azotobacter		Geotrichum	Vaucheria	
Bacillus		Glomerella		
Clostridium		Helminthosporium		
Corynebacterium		Mucor		
Escherichia		Myrothecium		
Flavobacterium		Penicillium		
Klebsiella		Stachybotrys		
Micrococcus		Trichoderma		
Paracolobactrum		Xylaria		
Proteus				
Pseudomonas				
Rhizobium				
Sarcina				
Serratia				
Sporocytophaga				
Thiobacillus				
Xanthomonas				

0.1 p.p.m. of DDT. No other metabolites were found (KEIL and PRIESTER 1969).

Two coliform bacteria (*Escherichia coli* and *Aerobacter aerogenes*), commonly found in aqueous environments, were found to metabolize DDT to TDE (MENDEL and WALTON 1966). In this particular study these cultures were isolated from rat feces, and resulted in the indication that the normal florae of the gastrointestinal tract are the major source of TDE found in animals fed DDT, rather than the liver.

In a similar study using the above two plus a third facultatively anaerobic species (*Klebsiella pneumonia*), WEDEMEYER (1966) indicated *A. aerogenes* as the most effective in the reductive dechlorination of DDT to TDE. A detailed study revealed that reduced Fe(II) cytochrome oxidase was responsible for DDT dechlorination. WEDEMEYER proposed that this may possibly explain the persistence of DDT in aerobic soils.

Anaerobic conversion of DDT to TDE in soil was reported by GUENZI and BEARD (1967). DDT was added to soil and the mixture was incubated anaerobically for two and four weeks. DDT and seven possible metabolic products were separated by thin-layer chromatography. The DDT was dechlorinated by unidentified soil microorganisms to TDE and only traces of other degradation products were detected (Table XI) (GUENZI and BEARD 1967). No degradation of DDT was detected in sterile soil.

Table XI. *Degradation products recovered from soil treated with DDT and incubated anaerobically* (GUENZI and BEARD 1967)

Materials found	Recovery (μg.) after	
	2 weeks	4 weeks
DDA	0.37	0.51
p-Chlorobenzoic acid	0.24	0.59
Dicofol	0.15	0.61
4,4'-Dichlorobenzophenone	0.39	0.64
TDE	7.1	35.0
DDT	62.0	19.0
4,4'-Dichlorodiphenylmethane	0.09	0.03
DDE	0.19	0.25
Total	71.0	57.0

HILL and McCARTY (1967) conducted an extensive study of organo-chlorine insecticide degradation under anaerobic conditions. The pesti-cides were mixed continuously at 35°C. with biologically active anaero-bic digested wastewater sludge obtained from a sewage treatment plant. The anaerobic conditions were ideal in that an active culture of anaerobic, methane-producing and sulfate-reducing bacteria were present. The metabolism of most of the organochlorine insecticides was more rapid under anaerobic than under the corresponding aerobic conditions, with the exception of heptachlor epoxide and dieldrin, which were very stable in both conditions. Extractable metabolic prod-ucts were more common in the anaerobic than in the corresponding aerobic conditions. Under anaerobic conditions the order of increasing persistence or stability of the chlorinated insecticides was lindane, hep-tachlor, endrin, DDT, TDE, aldrin, heptachlor epoxide, and dieldrin. DDT was metabolized very rapidly to TDE under anaerobic condi-tions, but persisted as DDT under aerobic conditions of several mg./l. of dissolved oxygen. An increase from 20° to 35°C. produced no signifi-cant increases in degradation rates except for the anaerobic metabolism of lindane and the aerobic metabolism of DDT. These insecticides were also more strongly sorbed by algae than by bentonite clay. The implica-tions of this study are that DDT would be readily converted to TDE on lake bottoms due to rapid anaerobic conversion by ooze organisms. Also implied is that lindane, aldrin, DDT, TDE, heptachlor, and pos-sibly endrin would undergo a significant amount of degradation in anaerobic muds of algae pounds or in natural waterways which reach temperatures of 20°C. or warmer.

In the above study the term "degradation" was used in a broad sense to refer to any measurable chemical change in a pesticide under natural environmental conditions. Metabolites or degradation products other than DDT to TDE were not identified. The algae used were a mixed culture consisting primarily of *Vaucheria sessilis,* a filamentous species. The implication here is that algae eventually die and frequently settle to the bottom where they may decay, along with chlorinated insecticides, anaerobically.

Water from Clear Lake, California, bovine rumen fluid, and porphyrins under anaerobic conditions were all found to convert DDT to TDE (MISKUS *et al.* 1965). Lake water containing large amounts of plankton converted more DDT to TDE than did small amounts. The authors believed that variation in oxygen content of water in different areas of the lake and different rates of oxygen depletion on incubation, caused by different population levels of florae and fauna, may have contributed to the wide range of TDE levels found. Distilled water showed no conversion of DDT to TDE, true also in boiled rumen fluid. Porphyrins under the proper, anaerobic, reducing conditions can convert DDT to TDE, which the authors indicated as one possible mechanism to explain this conversion in many biological systems.

In culture solutions, most of the actinomycetes and filamentous fungi tested degraded PCNB; several actinomycetes dechlorinated DDT to TDE, but no microorganisms degraded dieldrin (Table XII)

Table XII. *Degradation of DDT, dieldrin, and PCNB by actinomycetes and fungi in culture solutions*

| Microorganisms | Extent of degradation [a] | | |
	DDT	Dieldrin	PCNB
Fungi			
Aspergillus niger	—	—	+
Fusarium solani f. phaseoli	—	—	+
Glomerella cingulata	—	—	+
Helminthosporium victoriae	—	—	+
Mucor ramannianus	—	—	+
Myrothecium verrucaria	—	—	+
Penicillium frequentans	—	—	++
Trichoderma viride	—	—	+
Actinomycetes			
Nocardia sp.	++	—	+
Streptomyces albus	—	—	—
S. antibioticus	+	—	++
S. aureofaciens	++	—	+++
S. cinnamoneus	++	—	+
S. griseus	—	—	+
S. lavendulae	—	—	++
S. venezuelae	—	—	+
S. viridochromogenes	++	—	+

[a] = no. detectable degradation, + = less than ten percent, ++ = ten to 25 percent, and +++ = 25 to 50 percent.

(CHACKO *et al.* 1966). *Streptomyces aureofaciens* degraded PCNB to pentachloroaniline. Degradation of the pesticides in culture occurred only during the active growth phase of the actinomycetes or fungi, and stopped completely when growth ceased.

Dieldrin has since been shown to break down under the influence of a soil organism (MATSUMURA *et al.* 1968). This insecticide, which is highly stable in the presence of most microorganisms, was shown to metabolize in the presence of *Pseudomonas* sp. (*Shell* 33) originally isolated from a soil sample from the dieldrin factory yards of the *Shell Chemical Company* near Denver, Colorado. There were five major metabolites, with three lesser ones isolated. Surprisingly, one of the metabolites was identified as aldrin, the precursor to dieldrin, preceding its epoxidation in aerobic soils.

Lindane was actively degraded by unidentified microflora in flooded Philippine rice soils, but not when the soils were sterilized (RAGHU and MacRAE 1966). A second application of lindane 55 days later disappeared more rapidly than the first. The alpha, beta, gamma, and delta isomers disappeared from these flooded soils in 70 to 90 days MacRAE *et al.* 1967), and were believed to be degraded more actively by anaerobic than aerobic soil microflora under these conditions.

Two algae (*Chlorella vulgaris* and *Chlamydomonas reinhardtii*) metabolize lindane by dehydrochlorination to pentachlorocyclohexene, a non toxic lindane metabolite (SWEENEY 1968).

MacRAE *et al.* (1969) have observed the rapid anaerobic degradation of lindane by *Clostridium sp.* to a short-lived metabolite detectable by electron capture gas chromatography, definitely identified as not being pentachlorocyclohexene. After 27 hours incubation 75 percent of the theoretical amount of chlorine in the lindane was released by the bacteria as chloride ion in the reaction mixture.

b) Organophosphate insecticides

Four soil microorganisms, frequently identified in fresh water, were grown in pure cultures under aseptic conditions and exposed to organophosphate insecticides in metabolism studies by AHMED and CASIDA (1958). They were a yeast (*Torulopsis utilis*), two bacteria (*Pseudomonas fluorescens* and *Thiobacillus thiooxidans*), and a green alga (*Chlorella pyrenoidosa*). Rate studies were made on hydrolysis and oxidation of organophosphates added to the cultures as emulsions at 1,000 p.p.m. The compounds used were parathion, ronnel, dimefox, schradan, phorate, American Cyanamid 12008 (the propylthio analog of phorate) and the sulfinyl and sulfonyl derivatives, and the phosphorothiolate forms of phorate and AC 12008.

Chlorella and *Torulopsis* reacted similarly with the phosphorothioates, in that these toxicants were rapidly absorbed by the organisms and slowly released from living and dead cells in culture. With both, the S-(alkylthio)-methyl derivatives were more rapidly hydrolyzed than the S-(alkylsulfonyl)-methyl, which was slightly faster than the S-(alkylsulfinyl)methyls. In every instance the phosphorodithioate sulfoxides were the most and the phosphorothiolate sulfides the least stable to metabolism by these microorganisms.

Torulopsis and *Chlorella* oxidized the sulfides to sulfoxides, but *Chlorella* more effectively oxidized the dithioates to the thiolates. Little oxidation occurred with parathion, ronnel, dimefox, and schradan exposed to *Chlorella.*

The two bacteria did not oxidize, but did effectively hydrolyze phorate. The *Thiobacillus* did not utilize the sulfur in phorate.

Inquiring into the reasons for varying mosquito larvicidal longevity in the field, YASUNO *et al.* (1965) examined the effects of selected microorganisms on both organochlorine and organophosphate insecticides. Fenitrothion (Sumithion) used as mosquito larvicide was effective at 0.01 p.p.m., but the activity was lost after a few days. Ronnel, diazinon, fenthion, dieldrin, lindane, and DDT held up nicely for eight days, while fenitrothion, parathion, and methyl parathion were rapidly deactivated in bacteria-polluted water.

Among the organisms isolated, *Bacillus subtilis* was highly active in degrading fenitrothion, parathion, methyl parathion, and ronnel. Malathion, dichlorvos, and diazinon were not as readily degraded. Bacteria-free filtrates and sterilized cultures were inactive, indicating the living bacteria as the agent.

The main metabolic product was obtained by reduction of the nitro to the amino group. Amino-derivatives of all three compounds were identified after exposure to *B. subtilis.* The pH normally found in mosquito breeding places did not directly affect the insecticides.

Aspergillus oryzae was partly effective, but *Proteus vulgaris, Penicillium islandicum, Saccharomyces cereviciae,* and *Candida albicans* showed no effect in degrading fenitrothion.

Of 16 species of bacteria isolated from mosquito breeding waters, most degraded methyl parathion and fenitrothion into nontoxic products in one percent peptone culture solutions (HIRAKOSO *et al.* 1968). The insecticides studied were methyl parathion, fenitrothion, fenthion, diazinon, and dichlorvos (Table XIII).

Synergistic degradation of C^{14} ring-labeled diazinon by two bacterial genera was reported by GUNNER and ZUKERMAN (1968). *Arthrobacter* sp., known to attack the side chain of diazinon, was unable completely to metabolize the ring portion of the molecule. Similar tests with *Streptomyces* also indicated by it did not metabolize the pyrimidinyl carbon to carbon dioxide. However, when both were incubated together up to 20 percent of the C^{14} was captured as $C^{14}O_2$, suggesting the synergistic relationship between these organisms in metabolizing the pyrimidinyl moiety of diazinon. The only products identified in this study were carbon dioxide and the pyrimidinyl and phosphoro-

Table XIII. *Bacteria isolated from mosquito breeding waters and respective organophosphate insecticides degraded* (HIRAKOSO *et al.* 1968)

Bacteria	Deactivated insecticides
Pseudomonas aeruginosa	Fenthion
Serratia plymuthica	Dichlorvos
S. plymuthica	Methyl parathion
Escherichia coli	Fenitrothion
Paracolobactrum aerogenoides	Fenitrothion
Escherichia aurescens	Fenitrothion
Achromobacter eurydice	Fenitrothion
Pseudomonas convexa	Methyl parathion
P. putide	Methyl parathion
Escherichia freundii	Methyl parathion, fenitrothion, dichlorvos
Pseudomonas fluorescens	Methyl parathion
Serratia kiliensis	Methyl parathion
Flavobacterium aquatile	Methyl parathion

thioate moieties. The significant aspect of this study is the apparent synergistic relationship between two microorganisms in the metabolism of an insecticide molecule which neither can achieve alone.

Chlorella pyrenoidosa proteose, an alga found in aquatic environments as well as in the soil, has been shown as the responsible agent for metabolizing parathion around the roots of bean plants which in turn translocated the sulfur-containing metabolite to the above-ground parts (MACKIEWICZ *et al.* 1969). The algal metabolites consisted of 66 percent aminoparathion and a second product bearing the P—S group and the benzene ring. A trace amount of p-nitrophenol was found in the extracts but no paraoxon or p-aminophenol.

c) Herbicides

Metabolism of the five principal groups of herbicides has been discussed by KEARNEY (1966), classified into the phenylureas (fenuron, monuron, diuron), phenylcarbamates (CIPC, CEPC), s-trizaines (simazine, atrazine), chlorinated aliphatic acids (dalapon, TCA), and phenoxyalkanoic acids (2,4-D, 2,4-DB, 2,4,5-T).

The phenylureas will support the growth of bacteria *Xanthomonas, Sarcina, Bacillus,* and *Pseudomonas* and the fungi *Penicillium* and *Aspergillus*. In the metabolism of the phenylureas the dealkylation of methyl groups precedes hydrolysis of the urea linkage. To achieve this, first one methyl then the other is removed, followed by hydrolysis of urea to aniline. The resulting metabolic products are aniline, carbon dioxide, and ammonia.

The phenylcarbamates are generally less persistent than phenylureas. CIPC is degraded readily by *Pseudomonas striata, Flavobacterium* sp., *Agrobacterium* sp., and *Achromobacter* sp. CEPC is decomposed by *Achromobacter* sp., and *Arthrobacter* sp. The enzymatic hydrolysis of CIPC by cell-free extracts of *P. striata* yields Cl-aniline, carbon dioxide, and isopropanol.

The *s*-triazines can be degraded by the fungus *Fusarium roseum*, especially atrazine. *Aspergillus fumigatus*, also a fungus, liberated $C^{14}O_2$ only from the side chain, but not from the ring. Side chain degradation occurs in culture solutions with the fungi *Fusarium roseum*, *Geotrichum* sp., *Trichoderma* sp., and *Penicillium* sp.

With the chlorinated aliphatic acids, soil persistence studies indicate that dalapon is rapidly degraded while TCA degrades more slowly. Soil microorganisms can dehalogenate and use the carbon for energy. At this point there were seven bacteria, five fungi, and two actinomycetes effective in decomposing dalapon. *Arthrobacter* sp. completely breaks down dalapon. An enzyme from these broken cells liberates the chlorine ion yielding pyruvic acid. TCA is dehalogenated by soil microorganisms, yet most isolates grow feebly on TCA as a sole source of carbon.

A great amount of work has been done on the phenoxyalkanoic acids or 2,4-D group. At least ten different organisms are reported to decompose 2,4-D. In these studies two pathways are described: (1) *beta*-oxidation of the alkanoic acid and (2) initial hydrolysis of the ether linkage between the ring and the side chain. Step (1) proceeds by sequential removal of two carbon fragments from the functional end of the alkanoic acid. The fate of the ring structure in soils has also been studied. Detection of 2,4-dichlorophenol, 4-chlorocatechol, and chloromuconic acid from either soil or pure culture studies suggests a sequence of reactions involving ring hydroxylation and cleavage and further metabolism of the open structure to carbon dioxide.

With the phenylcarbamates IPC, CIPC, and analogs, Kaufman (1966) reported on studies conducted in soil perfusion systems with mixed populations of soil microorganisms. He concluded that position, type, and number of halogen substituents are important factors affecting the microbial decomposition of both aliphatic and certain aromatic herbicides. The number of carbon atoms composing the aliphatic acid also is an important factor. The number of carbon atoms present in alkyl substituents of alkylphenylcarbamate pesticides affects their rate of decomposition. Degradation of *meta*-substituted chlorophenylcarbamates was more rapid than *ortho*-chloro-, *para*-chloro-, or dichloro-substituted phenylcarbamates.

C^{14}-labeled CIPC and 2,4-D were studied during their exposure to microbial degradation in aqueous solutions by Schwartz (1967). Under the conditions normally found in water supplies, these two materials were strongly resistant to chemical attack, and very little of each would adsorb onto suspended mineral solids. The microorganisms used here were obtained from the activated sludge unit and the effluent from the primary sedimentation basin of a water reclamation plant, though the organism counts or identities were not revealed. CIPC was definitely degraded by biological action. The addition of nutrient broth to the microbial systems increased greatly the rate at which the iso-propyl-carbon atoms were metabolized, but had no effect on the rate of degradation of the phenyl group. SCHWARTZ (1967) proposed a metabolic pathway for the microbial degradation of CIPC by hydroly-sis to 3-chloroaniline and isopropanol. The latter is completely meta-bolized. The 3-chloroaniline is modified to 4-chlorocatechol which is further degraded by way of the muconic acid-ketoadipic acid pathway typical for aromatic compounds.

The microbial breakdown of 2,4-D proceeded at a slow rate with only a small relative concentration degraded by a mixed microbial population in a dilute medium of salts. In solutions containing 0.1 and 1.0 mg./l. of 2,4-D, no more than 37 percent of the acetic acid moiety disappeared over a period of three to six months. The presence of large amounts of nutrient broth had no appreciable effect on the rate of de-composition.

On this basis the author concluded that 2,4-D would not be de-graded materially by the microorganisms in a natural water supply. CIPC was considerably less resistant to biodegradation than 2,4-D, but may persist for weeks or months in a natural water environment. CIPC and 2,4-D have relatively short soil lives, but metabolism in water environments would be much slower than in soils. One might ex-pect that the longer soil-residual chlorinated insecticides, such as diel-drin, endrin, DDT, and BHC, which remain in soil for several years, once reaching a water supply may persist for extremely long periods of time in the absence of microorganisms.

DeMarco et al. (1967) studied the behavior of 2,4-D at 50 p.p.b. in simulated stratified impoundments, using a natural river water. This herbicide was biologically degraded under warm-aerobic (22° to 26° C., and two mg. oxygen/l.), cold-aerobic (10° to 13°C, and two mg. oxygen/l.), and cold deoxygenated or anaerobic (10°to 13°C, <0.3 mg. oxygen/l.). The low temperature, as might be expected, reduced the rate of biodegradation. Low oxygen concentration reduced the degradation rate of 2,4-D more than temperature, while the combina-tion of low temperature and low dissolved oxygen resulted in the least alteration, requiring 55 to 80 days to disappear. The warm-aerobic con-ditions accomplished the same task in six days, or a nine-fold increase in degradation rate. Total bacterial counts were made throughout the test, though not presented in the data or identified. These results indi-cate that certain environments present in impoundments could in-hibit the degradation of certain pesticides.

V. Discussion

Contrasts in interpretation of toxicity data are apparent when com-parisons are made of the data in Table V on the acute toxicity of DDT for *Daphnia pulex* (48 hr. LC_{50} of 0.4 p.p.m.) and the comments cited earlier from BUTLER and SPRINGER (1963) regarding the lack of adverse effect on growth of marine planktonic larvae in experiments involving three to six months exposures to DDT concentrations of one p.p.m. The reviewers recognize that identical species and test conditions were not involved; however, the situation reported by BUTLER and SPRINGER (1963) affords a more realistic basis for appraisal than data from acute toxicity tests.

Further basis for a conservative evaluation of acute data are found by contrasting the data concerning DDT (one p.p.m.) toxicity to estuarine phytoplankton, Table II, and those in Table III. The data in Table II indicate a decrease of 77 percent in carbon fixation while data in Table III for DDT at 0.1 p.p.m. indicate a mean decrease of 74 percent in C^{14} uptake. From these two incomparable values it is easy to conclude that a ten-fold increase in concentration produces very little increase in toxicity.

The complexity of ecosystems and the near simultaneous variety of man-induced environmental insults presents the scientist with a most difficult situation for evaluation. The complexities of the value judgments for the general public when confronted with diverse and often conflicting scientific conclusions are overwhelming.

The responses of microorganisms to and their reactions with pesticide components of their environment are a function not only of gross environmental dose and biochemical specificities of the microorganism but are also involved with size, shape, and gross composition of these life forms.

Surface area apparently plays an important role in the absorption of pesticides from water by microorganisms. From Table XIV (LA-MANNA and MALLETTE 1965) it can be seen that one g. of yeast contains 8.3 x 10⁹ particles and has 9,100 sq. cm. of surface area. Greater still is the enterobacterium, *Escherichia coli*, which has 1.8 x 10¹² particles/g. yielding 56,000 sq. cm. of surface. The very high surface:mass ratio for microorganisms, compared to other aquatic or terrestrial plants and animals, in part explains their characteristic rapid and somewhat thorough absorption of pesticides from the aquatic environment.

Another point to be considered is that most pesticides, particularly

Table XIV. *Comparison of the approximate dimensions of some microorganisms* (LAMANNA and MALLETTE 1965)

Microorganism	Radius (μ)	Volume (μ^3)	Surface (μ^2)	Surface — volume	No. particles/g.	Surface area/g. (cm.²)
Saccharomyces cereviciae	3	110	110	1	8.3 x 10⁹	9.1 x 10³
Escherichia coli	0.5	0.52	3.1	6	1.8 x 10¹²	5.6 x 10⁴
E. coli phage	0.004	2.5 x 10⁻⁴	0.02	80	3 x 10¹⁵	6 x 10⁵

insecticides, are lipophylic, resulting in a selective partitioning into or onto a large surface area containing surface lipids. The ether- or chloro- form-extractable lipids in microorganisms are shown in Table XV (PORTER 1946) indicating a range from 1.48 percent to 22.9 percent, depending on the organism and medium. These small forms contain all classes of lipids and lipoproteins, which are the main constituents of cell membranes and of certain intracellular structures such as mito- chondria and chloroplasts.

Because of the recent appearance of the polychlorobiphenyls (PCB's) in the analytical literature as contaminants which may be mistakenly identified as some insecticide residues, it is not reasonable to assume that all residues of DDT and related metabolites reported in the foregoing papers are in fact as stated. The PCB's are well recog- nized for their false identities as DDT and other organochlorine insecti- cides and their derivatives by electron capture gas chromatography (LICHTENSTEIN et al. 1969, RISEBROUGH et al. 1968, REICHEL et al. 1969, RISEBROUGH 1969, REYNOLDS 1969. SCHECHTER (1969) is of the opinion that the PCB's are primarily a problem in analyzing samples from aquatic rather than terrestrial environments. COON (1969) reported finding the PCB's more in marine birds than in other samples. Thus it would appear to the reviewers that the residues of DDT reported from marine and estuarine conditions deserve cautious acceptance because of their potential mistaken identity with the PCB's

Table XV. *Lipid content of various microorganisms* (PORTER 1946)

Microorganism	% Lipid (dry wt.)
Ether extractable [a]	
Corynebacterium	4.9
Bacillus	4.4
Escherichia coli	3.6–7.9
Yeast	5.0
Oospora lactis	7.5–22.5
Aspergillus	2.6–13.0
Penicillium	4.13–22.9
Mucor	7.03
Chloroform extractable	
Bacillus	1.48
Escherichia	11.77
Klebsiella	7.36
Proteus vulgaris	7.10
Pseudomonas	10.67
Yeast	2.92

[a] Ether, petroleum ether, or alcohol-ether extractable.

Table XVI. *Chemical designations of pesticide chemicals mentioned in text*

Pesticide	Chemical designation
Aldrin	1,4:5,8-Dimethanonaphthalene, 1,2,3,4,10,10-hexachloro-1,4,4a,5,8,8a-hexahydro-, *endo-exo* isomer
ASP-51	tetra-*n*-propyl dithiopyrophosphate
Atrazine	*s*-Triazine, 2-chloro-4-(ethylamino)-6-(isopropylamino)-
Azinphosmethyl	Phosphorodithioic acid, *O,O*-dimethyl ester, S-ester with 3-(mercaptomethyl)-1,2,3-benzotriazine-4(3*H*)-one
Bayer 37344	Carbamic acid, methyl-, 4-(methylthio)-3,5-xylyl ester
Baytex	Phosphorothioic acid, *O,O*-dimethyl *O*-[4-(methylthio)-*m*-tolyl] ester
BHC	Cyclohexane, 1,2,3,4,5,6-hexachloro-
Carbaryl	Carbamic acid, methyl-, 1-naphthyl ester
CEPC	Carbamic acid, *N*-(3-chlorophenyl)-1-chloro-2-propyl
Chlordane	4:7-Methanoindan, 1,2,4,5,6,7,8,8a-octachloro-3a,4,7,7a-tetrahydro-, *endo*-isomer
CIPC	Carbanilic acid, *m*-chloro-, isopropyl ester
2,4-D	Acetic acid, 2-4-dichlorophenoxy-
Dacthal	Terephthalic acid, tetrachloro, dimethyl ester
Dalapon	Propionic acid, 2,2-dichloro-
2,4-DB	Butyric acid, 4-(2,4-dichlorophenoxy)-
DDA	Acetic acid, bis(*p*-chlorophenyl)-
DDE	Ethylene, 1,1-dichloro-2,2-bis(*p*-chlorophenyl)-
DDT	Ethane, 1,1,1-trichloro-2,2-bis(*p*-chlorophenyl)-
DEF	Butyl phosphorotrithioate
Demeton	Phosphorothioic acid, *O,O*-diethyl-*O*-[2-(ethylthio)ethyl] ester mixed with *O,O*-diethyl-S-[2-(ethylthio)ethyl] ester
Diazinon	Phosphorothioic acid, *O,O*-diethyl *O*-(2-isopropyl-6-methyl-4-pyrimidinyl) ester
Dibrom	Phosphoric acid, 1,2-dibromo-2,2-dichloroethyl-, dimethyl ester
Dichlobenil	Benzonitrile, 2,6-dichloro-
Dichlorvos	Phosphoric acid, 2,2-dichlorovinyl-, dimethyl ester
Dicofol	Benzhydrol, 4,4'-dichloro-alpha-(trichloromethyl)-
Dieldrin	1,4:5,8-Dimethanonaphthalene, 1,2,3,4,10,10-hexachloro-6,7-epoxy-1,4,4a,5,6,7,8,8a-octahydro-, *endo-exo* isomer
Dimefox	Phosphorodiamidic fluoride, tetramethyl
Dipterex	See Trichlorfon
Diquat	Dipyrido[1,2-*a*:2',1'-*c*]pyrazidiinium compounds, "6,7-dihydro___. . ." dibromide
Di-Syston	Phosphorodithioic acid, *O,O*-diethyl S-[2-ethylthio)ethyl] ester
Diuron	Urea, 3-(3,4-dichlorophenyl)-1,1-dimethyl-
Dyrene	*s*-Triazine, 2,4-dichloro-6-(*o*-chloroanilino)-
Endothall	7-Oxabicyclo(2.2.1)heptane-2,3-dicarboxylic acid, disodium salt
Endrin	1,4:5,8-Dimethanonaphthalene, 1,2,3,4,10,10-hexachloro-6,7-epoxy-1,4,4a,5,6,7,8,8a-octahydro-, *endo-endo* isomer
Eptam	Carbamic acid, dipropylthio, S-ethyl ester
Ethion	Ethyl methylene phosphorodithioate
Fenac	Acetic acid, 2,3,6-trichlorophenyl-
Fenitrothion	Phosphorothioic acid, *O,O*-dimethyl *O*-4-nitro-*m*-tolyl ester

Table XVI. (continued)

Pesticide	Chemical designation
Fenthion	Phosphorothioic acid, O,O-dimethyl O-[4-(methythio)-m-tolyl] ester
Fenuron	Urea, 1,1-dimethyl-3-phenyl
Ferbam	Iron, tris(dimethyldithiocarbamato)-
Heptachlor	4:7-Methanoindene, 1,4,5,6,7,8,8a-heptachloro-3a,4,7,7a-tetrahydro-, $endo$-isomer
Heptachlor epoxide	4:7-Methanoindan, 1,4,5,6,7,8,8a-heptachloro-2,3-epoxy-3a,4,7,7a,tetrahydro-, $endo$-isomer
Hydram (Molinate)	S-ethyl hexahydro-1H-azepine-1-carbothioate
Imidan	Phosphorodithioic acid, O,O-dimethyl ester, S-ester with N-(mercaptomethyl)phthalimide
IPC	Carbamic acid, isopropyl-, N-phenyl ester
Kepone	1,3,4-Metheno-2H-cyclobuta[cd]pentalen-2-one, decachlorooctahydro-
Lignasan	ethylmercury phosphate
Lindane	Cyclohexane, 1,2,3,4,5,6-hexachloro-, $gamma$-isomer
Malathion	Succinic acid, mercapto-diethyl ester, S-ester with O,O-dimethyl phosphoro-dithioate
MCP	Acetic acid, 4-chloro-o-tolyloxy-
Metasystox	Phosphorothioic acid, O-[2-(ethylthio)-ethyl] O,O-dimethyl ester, mixed with S-[2-(ethylthio)ethyl] O,O-dimethyl ester
Methoxychlor	Ethane, 1,1,1-trichloro-2,2-bis(p-methoxyphenyl)-
Methyl parathion	Phosphorothioic acid, O,O-dimethyl O-p-nitrophenyl ester
Methyl Trithion	Phosphorodithioic acid, S-[(p-chlorophenyl)thiomethyl]-, O,O-dimethyl ester
Mirex	1,3,4-Metheno-2H-cyclobuta[cd]pentalene, dodecachlorooctahydro-
Monuron	Urea, 1,1-dimethyl-3-(p-chlorophenyl)-
Nabam	Carbamic acid, ethylene bisdithio-, disodium salt
Neburon	Urea, 1-butyl-3-(3,4-dichlorophenyl)-1-methyl-
N-Serve	2-chloro-6-(trichloromethyl) pyridine
Paraquat	Bipyridinium compounds, 1,1-dimethyl-4,4′_____ . . . dimethyl sulfate or dichloride
Parathion	Phosphorothioic acid, O,O-diethyl O-p-nitrophenyl ester
PCNB	Benzene, pentachloronitro-
Phaltan	Phthalimide, N-[(trichloromethyl)thio]-
Phorate	Phosphorodithioic acid, O,O-diethyl S-[(ethylthio)methyl] ester
Ronnel	Phosphorothioic acid, O,O-dimethyl O-2,4,5-trichloro-phenyl ester
Schradan	Pyrophosphoramide, octamethyl-
Silvex	Propionic acid, 2-(2,4,5-trichlorophenoxy)-
Simazine	s-Triazine, 2-chloro-4,6-bis(ethylamino)-
2,4,5-T	Acetic acid, 2,4,5-trichlorophenoxy-
TCA	Acetic acid, trichloro-
TDE	Ethane, 1,1-dichloro-2,2-bis(p-chlorophenyl)-
TEPP	Ethyl pyrophosphates
Thiodan	5-Norbornene-2,3-dimethanol, 1,4,5,6,7,7-hexachloro-, cyclic sulfite
Tillam	Carbamic acid, butylethylthio-S-propyl ester

Table XVI. (continued)

Pesticide	Chemical designation
Tordon	4-Amino-3,5,6-trichloropicolinic acid
Toxaphene	Chlorinated camphene containing 67 to 69 percent chlorine
Trichlorfon	Phosphonic acid, (2,2,2-trichloro-1-hydroxyethyl)-, dimethyl ester
Trifluran	Trifluoro-2,6-dinitro-N,N,-dipropyl-*p*-toluidine
Vernam	Carbamic acid, dipropylthio-, *S*-propyl ester
Zytron	Phosphoroamidothioic acid, isopropyl-, *O*-(2,4-dichloro-phenyl)-, *O*-methyl ester

Summary

Pesticides do not always interact with aquatic microorganisms as predicted. Generally all pesticides are toxic to all microorganisms at some dosage, the adage "the poison is in the dosage" holding true. Toxicity as measured in the several reported studies includes changes in growth rate, metabolic rate, and photosynthesis.

The phenylureas are the most toxic herbicides to phytoplankton, while surprisingly the cyclodienes are the most toxic insecticides. DDT can reduce photosynthesis in phytoplankton, and is also the most toxic material to many crustaceae.

Aquatic microorganisms absorb and concentrate pesticides from water apparently inversely related to the water solubility of the compound, DDT being the notable object of numerous studies. Living organisms do not seem to be any more efficient than dead organisms in this seemingly nonspecific, physical property of microorganisms.

Metabolism of pesticides in microorganisms is as varied as in vertebrates. Again it appears that any of the small forms will metabolize any of the pesticides to some extent, perhaps with the exception of dieldrin. In the case of DDT there are probably two routes of metabolism: aerobic, leading to the formation of DDE, and anaerobic, which produces TDE. In the case of actinomycetes metabolism occurred only during the active growth phase stopping completely when growth ceased. Phosphate insecticides are readily metabolized by all bacteria, actinomycetes, and fungi and algae examined. The five classes of herbicides are probably attacked by all forms, some being highly selective.

Résumé *

Intéraction des pesticides avec les microorganismes aquatiques et le plancton

Les pesticides ne réagissent pas toujours avec les microorganismes aquatiques comme prévu. En général, tous les pesticides sont toxiques

---

* Traduit par R. MESTRES.

pour tous les microorganismes à un certain dosage, l'adage: "la dose crée le poison" étant vrai. Plusieurs études citées montrent que la toxicité se traduit par des changements dans la vitesse de croissance, le métabolisme et la photosynthèse.

Les desherbants les plus toxiques pour le plancton sont les phényl-urées et il est surprenant que les cyclodiènes soient les insecticides les plus toxiques. Le zeidane peut diminuer la photosynthèse dans le phytoplancton et est également le composé le plus toxique pour de nombreux crustacés.

Les microorganismes aquatiques paraissent absorber et concentrer les pesticides de l'eau en fonction inverse de la solubilité dans l'eau du composé, le zeidane ayant fait l'objet de nombreuses études. Les organismes vivants ne semblent pas être plus efficaces que les morts dans cette propriété physique apparemment non spécifique des microorganismes.

Le métabolisme des pesticides dans les microorganismes est aussi varié que chez les vertébrés. Il apparaît encore que tout petit individu métabolisera plus ou moins un pesticide quelconque, à l'exception peut-être de la dieldrine. Dans le cas du zeidane, il existe probablement deux voies métaboliques, l'une aérobie conduisant à la formation de DDE et l'autre anaérobie qui forme le DDD. Avec les actinomycètes le métabolisme se produit seulement au cours de la phase active de croissance et s'arrête complètement lorsque la croissance cesse. Les insecticides organophosphorés sont rapidement métabolisés par toutes les bactéries, actinomycètes, champignons et algues étudiées. Les cinq catégories de desherbants sont probablement attaquées par tous les microorganismes, certains étant hautement sélectifs.

Zusammenfassung *

Wechselwirkung von Pestiziden mit aquatischen Mikroorganismen und mit Plankton

Pestizide beeinflussen Mikroorganismen in wässerigem Milieu nicht immer in der vorausgesagten Weise. Im Allgemeinen sind alle Pestizide in gewissen Dosen für alle Mikroorganismen toxisch, sobaß die Redewendung "das Gift liegt in der Dosierung" zu recht besteht. Die Toxizität wird in den verschiedenen referierten Untersuchungen als Änderung der Wachstumsrate, der Stoffwechselintensität und der Photosynthese gemessen.

Die Phenylharnstoffe sind die am meisten toxisch befundenen Herbizide für das Phytoplankton, während überraschenderweise die Cyklodiene die am meisten toxischen Insektizide darstellen. DDT kann die Photosynthese beim Phytoplankton reduzieren und ist ebenfalls für zahlreiche Crustaceen das am stärksten toxische Material.

---

* Übersetzt von H. F. Linskens.

Die im Wasser lebenden Mikroorganismen absorbieren und konzentrieren Pestizide offensichtlich aus dem Wasser im umgekehrten Verhältnis zur Wasserlöslichkeit der Verbindungen, wobei DDT das häufigste Objekt der zahlreichen Untersuchungen ist. Lebende Organismen scheinen dabei nicht wirksamer zu sein, als tote. Es handelt sich also wahrscheinlich um eine unspezifische, physikalische Eigenschaft der Mikroorganismen.

Der Metabolismus der Pestizide in Mikroorganismen ist ebenso vielfältig wie bei den Vertebraten. Wiederum scheint es so, als ob alle kleinen Formen jedes Pestizid bis zu einem gewissen Grade in den Stoffwechsel einbeziehen, vielleicht mit Ausnahme von Dieldrin. Im Falle des DDT liegen wahrscheinlich zwei Stoffwechselwege vor: ein aerober, der zur Bildung von DDE führt, sowie ein anaerober, der zur Bildung von DDD führt. Im Falle der Aktinomyceten erfolgt die Metabolisierung ausschließlich während der aktiven Wachstumsphase und wird vollständig gestoppt, sobald das Wachstum aufhört. Phosphor-Insektizide werden von allen Bakterien, Aktinomyceten und Pilzen, sowie den untersuchten Algen rasch metabolisiert. Die fünf Klassen von Herbiziden werden wahrscheinlich von allen Formen abgebaut, einige mit hoher Selektivität.

References

Ahmed, M. K., and J. E. Casida: Metabolism of some organophosphate insecticides by microorganisms. J. Econ. Entomol. **51**, 59 (1958).

Alexander, M.: Microbiology of pesticides and related hydrocarbons. In: Principles and applications in aquatic microbiology. Proc. Rudolfs Research Conf. Rutgers, N.J. New York: Wiley (1964).

Anonymous: Pesticide-wildlife studies. *U.S. Department of Interior*, Fish and Wildlife Service, Circ. 167 (1963).

Audus, L. J., Ed.: The physiology and biochemistry of herbicides. London and New York: Academic Press (1964).

Butler, P. A.: Effects of herbicides on estuarine fauna. S. Weed Control Conf. Proc. **18**, 576 (1965a).

—— Commercial fishery investigations. *U.S. Department of Interior*, Fish and Wildlife Service, Circ. 226, p. 65 (1965 b).

—— Pesticides in the estuary. Proc. Marsh. Estuary Mgt. Symp., p. 120 (1967).

——, and P. F. Springer: Pesticides—A new factor in coastal environments. Trans. 28th N. Amer. Wildlife Conf., p. 378 (1963).

Cabejszek, I., and J. Stanislawska: Effect of methyl parathion (p-nitrophenyl O,O-dimethyl thionophosphate) on water-borne organisms. Roczniki Panstwowego Zakladu Higieny **17**, 353 (1966).

—— —— Effects of thometon (O,O-dimethylthio-phosphate 2-ethyl mercaptoethyl) on water organisms. Roczniki Panstwowego Zakladu Higieny **18**, 155 (1967).

Chacko, C. I., and J. L. Lockwood: Accumulation of DDT and dieldrin by microorganisms. Can. J. Microbiol. **13**, 1123 (1967).

—— ——, and M. Zabik: Chlorinated hydrocarbon pesticides: Degradation by microbes. Science **154**, 893 (1966).

Coon, F. B.: Private communication (1969).

COPE, O. B.: Contamination of the freshwater ecosystem by pesticides. J. Applied Ecol. 3 (Suppl.), 33 (1966).

COWELL, B. C.: The effects of sodium arsenite and silvex on the plankton populations in farm ponds. Amer. Fish. Soc. Trans. 98, 371 (1965).

CRANCE, J. H.: The effects of copper sulfate on Microcystis and zooplankton in ponds. Progr. Fish-Culturist 25, 198 (1963).

DE MARCO, J., J. M. SYMONS, and G. G. ROBECK: Behavior of synthetic organics in stratified impoundments. Amer. Water Works Assoc. J. 59, 965 (1967).

EDWARDS, C. A.: Insecticide residues in soil. Residue Reviews 13, 83 (1966).

FROBISHER, M., JR.: Fundamentals of bacteriology, 4th ed. Philadelphia: W. B. Saunders Co. (1949).

FUNDERBURK, H. H., JR., and G. A. BOZARTH: Review of the metabolism and decomposition of Diquat and Paraquat. J. Agr. Food Chem. 15, 563 (1967).

GOTTLIEB, D.: The disappearance of antibiotics from soil. Abstr., Phytopathol. 42, 9 (1952).

GREGORY, W. W., JR., J. K. REED, and L. E. PRIESTER, JR.: Accumulation of parathion and DDT by some algae and protozoa. J. Protozool. 16, 69 (1969).

GUENZI, W. D., and W. E. BEARD: Anaerobic biodegradation of DDT to DDD in soil. Science 156, 1116 (1967).

GUNNER, H. B., and B. M. ZUCKERMAN: Degradation of 'Diazinon' by synergistic microbial action. Nature 217, 1183 (1968).

GUNTHER, F. A., W. E. WESTLAKE, and P. S. JAGLAN: Reported solubilities of 738 pesticide chemicals in water. Residue Reviews 20, 1 (1968).

HARDY, J. L.: Effect of tordon herbicides on aquatic chain organisms. Down to Earth 22, 11 (1966).

HICKEY, J. J., J. A. KEITH, and F. B. COON: An exploration of pesticides in a Lake Michigan ecosystem. J. Applied Ecol. 3 (Suppl.) 141 (1966).

HILL, D. W., and P. L. MCCARTY: Anaerobic degradation of selected chlorinated hydrocarbon pesticides: J. Water Pollution Contr. Fed. 39, 1259 (1967).

HIRAKOSO, S., I. KITAGO, and C. HARINASUTA: Inactivation of insecticides by bacteria isolated from polluted waters where the mosquito larvae breed in large number. Med. J. Malaya 22, 249 (1968).

JONES, B. R., and J. B. MOYLE: Population of plankton animals and residual chlorinated hydrocarbons in soils of six Minnesota ponds treated for control of mosquito larvae. Trans. Amer. Fish. Soc. 92, 3 and 121 (1963).

KALLMAN, B. J., and A. K. ANDREWS: Reductive dechlorination of DDT to DDD by yeast. Science 141, 1050 (1963).

KASAHARA, S.: Studies on the biology of the parasitic c pepod, Lernaea cyrinacea Linnaeus, and the methods of controlling this parasite in fish culture. Contr. Fish. Lab., Faculty of Agr., Univ. of Tokyo 3, 103 (1962).

KAUFMAN, D. D.: Structure of pesticides and decomposition by soil microorganisms. In: Pesticides and their effects on soils and water. Amer. Soc. Agron. Special Publ. No. 8. Symposium papers sponsored by Soil Sci. Soc. Amer. (1966).

KEARNEY, P. C.: Metabolism of herbicides in soils. In: Organic pesticides in the environment. Adv. Chem. Series 60, 250 (1966).

KEIL, J. E., and L. E. PRIESTER: DDT uptake and metabolism by a marine diatom. Bull. Environ. Contamination Toxicol. 4, 169 (1969).

KO, W. H., and J. L. LOCKWOOD. Accumulation and concentration of chlorinated hydrocarbon pesticides by microorganisms in soil. Can. J. Microbiol. 14, 1075 (1968).

LAMANNA, C., and M. F. MALLETTE: Basic bacteriology, ref. p. 61. Baltimore: Williams and Wilkins (1965).

LAWRENCE, J. M.: Aquatic herbicide data. Agr. Handbook No. 231 (1962).

LAZAROFF, N.: Algal response to pesticide pollutants. Bacteriol. Proc. 48, G149 (1967).

LICHTENSTEIN, E. P., K. R. SCHULZ, T. W. FUHREMANN, and T. T. LIANG: Biological interaction between plasticizers and insecticides. J. Econ. Entomol. **62**, 761 (1969).

—— ——, R. F. SKRENTNY, and Y. TSUKANO: Toxicity and fate of insecticide residues in water. Arch. Environ. Health **12**, 199 (1966).

LUCZAK, J., and J. MALESZEWSKA: Effect of Thiometon (O,O-dimethylthiophosphate 2-ethyl mercaptoethyl) on physio-chemical properties and development of bacteria in water. Roczniki Panstwowego Zakladu Higieny **18**, 151 (1967).

MACKIEWICZ, M., K. H. DEUBERT, H. B. GUNNER, and B. M. ZUCKERMAN: Study of parathion biodegradation using gnotobiotic techniques. J. Agr. Food Chem. **17**, 129 (1969).

MARTIN, J. P.: Influence of pesticides on soil microbes and soil properties. In: Pesticides and their effects on soils and water. Amer. Soc. Agron. Special Publ. No. 8, p. 95. Symposium papers sponsored by Soil Sci. Soc. Amer. (1966).

MATSUMURA, F., G. M. BOUSH, and A. TAI: Breakdown of dieldrin in the soil by a microorganism. Nature **219**, 965 (1968).

MACRAE, I. C., and M. ALEXANDER: Microbial degradation of selected herbicides in soil. J. Agr. Food Chem. **13**, 72 (1965).

——, K. RAGHU, and E. M. BAUTISTA: Anaerobic degradation of the insecticide lindane by Clostridium sp. Nature **221**, 859 (1969).

—— ——, and T. F. CASTRO: Persistence and biodegradation of four common isomers of benzene hexachloride in submerged soils. J. Agr. Food Chem. **15**, 911 (1967).

MENDEL, J. L., and M. S. WALTON: Conversion of p,p'-DDT to p,p'-DDD by intestinal flora of the rat. Science **151**, 1527 (1966).

MISKUS, R. P., D. P. BLAIR, and J. E. CASIDA: Conversion of DDT to DDD by bovine rumen fluid, lake water, and reduced porphyrins. J. Agr. Food Chem. **13**, 481 (1965).

MULLIGAN, H. F.: Management of aquatic vascular plants and algae. Internat. Symp. on Eutrophication, Madison, Wis. (1967).

MUNNECKE, D. E.: Fungicides. In D. C. Torgeson (ed.), Vol. 1. New York: Academic Press (1966).

ODUM, W. E., G. M. WOODWELL, and C. F. WURSTER: DDT residues absorbed from organic detritus by fiddler crabs. Science **164**, 576 (1969).

PIERCE, MADELENE: The effect of the weedicide Kuron upon the flora and fauna of two experimental areas of Long Pond, Dutchess County, N.Y. N.E. Weed Control Conf. Proc. **12**, 338 (1958).

—— Progress report of the effect of Kuron upon the biota of Long Pond, Dutchess County, N.Y. N.E. Weed Control Conf. Proc. **14**, 472 (1960).

PORTER, J. R.: Bacterial chemistry and physiology, p. 407. New York: Wiley (1946).

PRAMER, D.: The persistence and biolgcial effects of antibiotics in soil. Applied Microbiol. **6**, 221 (1958).

RAGHU, K., and I. C. MACRAE: Biodegradation of lindane in submerged soils. Science **154**, 263 (1966).

—— —— The effect of the gamma-isomer of BHC upon the microflora of submerged rice soil. I. Effect upon algae. Can. J. Microbiol. **13**, 173 (1967 a).

—— —— The effect of the gamma-isomer of BHC upon the microflora of submerged rice soil. II. Effect upon nitrogen mineralisation and fixation and selected bacteria. Can. J. Microbiol. **13**, 625 (1967 b).

REICHEL, W. L., T. G. LAMONG, E. CROMARTIE, and L. N. LOCKE: Residues in two bald eagles suspected of pesticide poisoning. Bull. Environ. Contamination Toxicol. **4**, 24 (1969).

REYNOLDS, L. M.: Polychlorobiphenyls (PCB's) and their interference with pesticide residue analysis. Bull. Environ. Contamination Toxicol. **4**, 128 (1969).

RISEBROUGH, R. W., D. B. PEAKALL, S. G. GERMAN, M. N. KIRVEN, and P. REICHE: Polychlorinated biphenyls in the global system. Nature 220, 1098 (1968).
——, P. REICHE, and H. S. OLCOTT: Current progress in the determination of the polychlorinated biphenyls. Bull. Environ. Contamination Toxicol. 4, 192 (1969).
ROBERTS, J. E., R. D. CHISHOLM, and L. KOTLITSY: Persistence of insecticides in soil. J. Econ. Entomol. 55, 153 (1962).
ROBINSON, J., A. RICHARDSON, A. N. CRABTREE, J. C. COULSON, and G. R. POTTS: Organochlorine residues in marine organisms. Nature 214, 1307 (1967).
ROGOFF, M. H.: Oxidation of aromatic compounds by soil bacteria. Adv. Applied Microbiol. 3, 193 (1961).
SCHECTER, M. S.: Private communication (1969).
SCHWARTZ, H. G., JR.: Microbial degradation of pesticides in aqueous solutions. J. Water Pollution Control Fed. 39, 1701 (1967).
SEAMAN, D. E., and T. M. THOMAS: Absorption of herbicides by submersed aquatic plants. Proc. Calif. Weed Conf., p. 11 (1966).
SETHUNATHAN, N., and I. C. MacRAE: Some effects of diazinon on the microflora of submerged soils. Plant and Soil 30, 109 (1969).
SWEENEY, R. A.: Metabolism of lindane by unicellular algae. Proc. 12th Conf. Great Lakes Research (1968).
TATUM, W. M., and R. D. BLACKBURN: Preliminary study of the effects of diquat on the natural bottom fauna and plankton in two subtropical ponds. S.E. Assoc. Game & Fish Comm. Proc. Ann. Conf., p. 16 (1962).
THIEGS, B. J.: Microbial decomposition of herbicides. Down to Earth (Dow Chemical Co.), Fall issue, p. 7 (1962).
TOTH, S. J., and D. N. RIEMER: Precise chemical control of algae in ponds. J. Amer. Water Works Assoc. 60, 367 (1968).
UKELES, R.: Growth of pure cultures of marine phytoplankton in the presence of toxicants. Applied Microbiol. 10, 532 (1962).
WARE, G. W., M. K. DEE, and W. P. CAHILL: Water florae as indicators of irrigation water contamination by DDT. Bull. Environ. Contamination Toxicol. 3, 333 (1968).
WATSON, G. H., and W. B. BOLLEN: Effect of copper sulfate weed treatment on bacteria in lake bottoms. Ecology 33, 522 (1952).
WEDEMEYER, G.: Dechlorination of DDT by Aerobacter aerogenes. Science 152, 647 (1966).
WESTLAKE, W. E., and F. A. GUNTHER: Organic pesticides in the environment. Adv. Chem. Series 60, 110 (1966).
WOODFORD, E. K., and G. R. SAGAR (Eds.): Herbicides and the soil. Oxford: Blackwell Scientific (1960).
WOODWELL, G. M., C. F. WURSTER, JR., and P. A. ISAACSON: DDT residues in an east coast estuary: A case of biological concentration of a persistent insecticide. Science 156, 821 (1967).
WURSTER, C. F., JR.: DDT reduces photosynthesis by marine phytoplankton. Science 159, 1474 (1968).
YASUNO, M., S. HIRAKOSO, M. SASA, and M. UCHIDA: Inactivation of some organophosphorous insecticides by bacteria in polluted water. Japan J. Expt. Med. 35, 545 (1965).

The photochemistry of halogenated herbicides

By

J. R. PLIMMER [*]

Contents

I. Scope of review

The scope of this review has been limited primarily to the photo-chemical loss of halogen from herbicides, although some discussion of competitive and subsequent photolysis of pathways of individual herbicides has been included. The common names [1] approved by the Weed Science Society of America (WEED SOCIETY OF AMERICA 1967) have been used.

---

[*] Crops Research Division, Agricultural Research Service, U.S. *Department of Agriculture,* Beltsville, Maryland. Mention of a trademark name or a proprietary product does not constitute a guarantee or warranty of the product by the U.S. Department of Agriculture and does not imply its approval to the exclusion of other products that may also be suitable.

[1] Common names of herbicides and their chemical designations are listed in Table II.

47

A recent review has appeared in which the photochemistry of the herbicides has been summarized (CROSBY and LI 1969). This review contains an excellent discussion of the alteration of herbicides by solar radiation, and provides a valuable introduction to the field. The reader is referred to this review for an account of experimental photochemical arrangements and radiation sources.

The aim of the present review is to summarize available knowledge of the photochemistry of the halogenated herbicides, and to discuss this in the light of present knowledge of the mechanism of photochemical loss of chlorine from aromatic systems. The introductory section of this review provides a brief outline of the light absorption process as it relates to the discussion. More detailed treatment is available in a number of recent texts (TURRO 1965, NECKERS 1967, CALVERT and PITTS 1966).

II. Photochemical reactions

Light is electromagnetic in nature. The propagation of light involves the transfer of energy. The energy transferred depends on the wavelength of the light and its intensity. The shorter the wavelength, the greater the energy of the radiation.

Radiation may be absorbed by a molecule, thereby increasing the energy of the molecule. Energy absorbed may increase the translational, rotational, vibrational, or electronic energy of the molecule. If the wavelength of the radiation imparts sufficient energy to interact with the valence electrons of the molecule, an electronically excited molecule results. Energy may be lost by the excited molecule in a number of ways: one of these is chemical reaction.

Light absorption precedes photochemical reaction. Many herbicides absorb light in the ultraviolet region of the spectrum; therefore ultraviolet sources are commonly used in the study of their photochemical reactions. Solar energy contains no appreciable component below 2950A (KOLLER 1966). It is therefore important in photochemical research to select the wavelengths used for irradiation. The products of a reaction are often wavelength dependent. The result of energy absorption may be homolytic bond fission. Consequently, the reactions of the initially produced radicals will depend on the physical state of the irradiated compound, the solvent, and on other reactants present, such as oxygen. The particular wavelengths of light absorbed (i.e., the absorption spectrum) by a substance will depend on its physical state, i.e., whether it is in the gas phase, in solution, adsorbed on a surface, or is irradiated as a pure solid. Alternatively a reaction may be photosensitized. The energy transfer may be carried out by the intermediacy of a molecule which absorbs light (a photosensitizer) and, by way of a relatively long-lived excited state, passes on energy to bring about the photolysis of another molecular species present in the medium.

Failure to react photochemically may result from the failure of a molecule to absorb light; alternatively, stable excited states may be formed which dissipate energy by fluorescence, phosphorescence, or other pathways.

If a molecule absorbs light and passes from an initial state of energy E_1, to a state of final energy E_2, the increase in energy

$$\Delta E = E_2 - E_1 = h\nu \qquad (i)$$

where h is Planck's constant, ν is the frequency of the absorbed radiation, and

$$\nu = c/\lambda, \qquad (ii)$$

where c is the velocity of light and λ is the wavelength. The shorter wavelengths therefore transmit greater energy. Quantitative aspects have been discussed by NECKERS (1967). If the appropriate numerical values are substituted in equations (i) and (ii) in molar terms, the absorption of one einstein of light (6.023×10^{23} quanta or one mole of light quanta) of wavelength 3,000 A will impart 94.6 kcal./mole to the absorbing molecule. This energy increment is of the same magnitude as many covalent bond energies, and is therefore adequate to bring about dissociation. As sufficient energy is available to effect dissociation of covalent bonds, photochemical reactions often involve free radicals and the products may frequently be predictable on this hypothesis.

The absorption of light takes place in discrete quanta. Each quantum absorbed activates one molecule in the primary excitation process. The primary processes may be defined as the processes involving the actual initiation of the photochemical reaction, or as the processes involved in the formation of an excited state due to direct absorption of a quantum of light. Secondary processes involve reaction of molecular fragments or the dissipation of energy as the molecule returns to the ground state.

The products of the photochemical reaction depend on the sequence of events following the excitation process. An important relationship is the quantum yield defined in the following equation:

$$\text{Quantum yield} = \frac{\text{No. of moles of resultant product}}{\text{No. of einsteins of light absorbed}} \qquad (iii)$$

The relationship among the efficiency of the photochemical reaction, the percentage yield, and reactivity has been discussed in a review by TURRO (1967). He concludes that the percentage yield of a photochemical reaction is only remotely related to the inherent reactivity of an electronically excited state. The rate constant for the reaction of an electronically excited state is the proper measure of reactivity in photochemical reactions. Some of the comparative results, discussed later in this review, should be considered with these reservations in mind.

III. Mechanism

Photochemical reactions of halogenated aromatic compounds appear to be free-radical in nature. In alcohol or benzene solution a radical generated by the fission of a carbon-halogen bond reacts with the solvent.

The photolysis of iodobenzene in the liquid state with a mercury vapor lamp gave a complex mixture of products. It was concluded from the composition of the mixture that free phenyl radicals were produced in the reaction (BLAIR and BRYCE-SMITH 1960). In dilute solutions of halogen compounds, the reactions were found to be of synthetic value. For example, irradiation of 2-iodophenol in benzene gave 2-hydroxybiphenyl in 60 percent yield. A variety of compounds in which iodine was replaced by phenyl were prepared by irradiation of the appropriate iodo compound (WOLF and KHARASCH 1965). Irradiation of 2-iodobenzoic acid in benzene gave 2-phenylbenzoic acid. The same product resulted from photolysis of o-chlorobenzoic acid in benzene (PLIMMER and HUMMER 1968 a). The reactions of chlorinated aromatics appear similar in mechanism and type to those of the iodo compounds; the difference in bond strengths will have some influence, however, and between iodine and chlorine derivatives there may be a difference in excited states from which reaction occurs. This is discussed in a later paragraph.

The free-radical character of the reaction leads to products which are solvent dependent. In benzene, phenylation is the principle reaction and a new C–C bond is formed. In alcohols, however, hydrogen abstraction is an important process and the photochemically generated free radical reacts to give a reduced product. Halide ion remains in solution and an increase of pH is commonly observed.

Chlorine loss probably proceeds by way of a singlet state.[2] Chlorobenzoic acid did not show appreciable photoreaction in the presence or absence of benzophenone during eight hours irradiation by light of wavelength > 2,850A (PLIMMER and HUMMER 1968 c). A mechanistic study of the photocyclization of halogenated phenylnaphthalenes demonstrated that photoreaction probably occurred via the singlet state of the bromo- and chloro-substituted compounds whereas the iodo derivative reacted via a triplet state (triplet energy, $E_t = 60$ kcal.) since the benzophenone ($E_t = 69$ kcal.) sensitized photolysis afforded the same products as direct excitation in the latter case alone. The bromo and chloro compounds ($E_t = 60$ kcal. for both) failed to afford products by photosensitization. The reaction under investigation was the photocyclization of a 1-o-halophenylnaphthalene (1) in benzene to yield fluoranthene (2). The iodo compound decomposed relatively

---

[2] For a discussion of excitation processes see TURRO (1965).

(1) (2)

rapidly affording no fluoranthene but fluoranthene was obtained as major product from the slower reaction of the chloro derivative. The behavior of the bromo compound was intermediate. The quantum yield of the photolytic reaction resulting in ring closure of chloro- and bromonaphthalene derivatives was not affected by benzophenone. A triplet excited state is therefore unlikely (HENDERSON and ZWEIG 1967).

The photochemical replacement of halogen by hydroxyl is formally analogous to a hydrolytic reaction. Indeed, the rate of replacement of chlorine by the hydroxyl group of chloroacetic acid is enhanced by ultraviolet light (EULER 1916). The replacement of halogen in 2-bromo (or chloro) -4-nitroanisole (3) irradiated in $0.1N$ sodium hydroxide/tetrahydrofuran solution by an ultraviolet source with a borosilicate glass filter occurred to give a 44 percent yield of 5-nitroguiacol (4). No hydroxylated product was obtained by irradiation of a 2- or 4-bromo-

(3) (4)

anisole (NIJHOFF and HAVINGA 1965). Photonucleophilic substitution reactions of a number of monosubstituted benzenes have been investigated, and low yields of product were obtained in high concentrations of nucleophile. The reactions were interpreted as proceeding by an ionic mechanism (BARLTROP et al. 1967).

Replacement of halogen by the hydroxyl group in a number of pesticides is reported as occurring in aqueous solution, usually with free access of air, under illumination from sunlight or ultraviolet light. The question of photosensitization by trace metals or other impurities and the role of oxygen in the reaction have not been studied. Reaction occurs slowly in a number of cases, although it seems unlikely that there is radiation present of sufficiently short wavelength to be directly absorbed by the reacting molecule. Irradiation of chlorobenzoic acids and amiben (3-amino-2, 5-dichlorobenzoic acid) in water with light of wavelength directly absorbed by the molecule leads to replacement of chlorine by hydrogen in addition to replacement by hydroxyl. Hydrogen abstraction from water molecules occurs in these reactions and hydroxylation is also observed (CROSBY 1966, PLIMMER and HUMMER 1968 a).

The photochemistry of substituted aromatic compounds has been studied in detail by flash-photolysis techniques. The initial products of photolysis of a number of aromatic compounds in aqueous solution were identified. Many easily oxidized aromatics initially generated a hydrated electron in aqueous solution and afforded radical products. Aromatics with electron-attracting substituents did not release electrons. Among these were halogen substituted derivatives which reacted by losing halogen through C–X bond fission. An aqueous solution of 4-bromophenol, irradiated at 2,537A, afforded 4,4'-dihydroxybiphenyl and hydroquinone; 3-bromophenol gave resorcinol and 3,3'-dihydroxybiphenyl; and 2-bromophenol gave catechol and 2,2'-dihydroxybiphenyl. The bromophenols did not generate a hydrated electron but reacted by fission of a C–Br bond which requires ~ 67 kcal./mole.

The following reaction scheme (Scheme 1) has been proposed for the irradiation process with initial formation of a radical pair. One or other of the radicals then reacts with water.

Brackets represent the solvent cage. Rearrangements as:

did not occur. No phenoxyl radical could be detected and coupling products were absent. Only one symmetrical dimer was produced for each individual bromophenol. The composition of the products of the photolysis depend on the solvent. Major changes are to be expected when the solvent is a reactant. Proton abstraction from cyclohexane is 20 kcal. easier than from water. There are sufficient examples of proton (rather than hydroxyl) abstraction from methanol to indicate that a relative difference in bond strengths is the underlying cause. The following bond dissociation energies in kcal./mole for methanol, water, benzene, and cyclohexane illustrate this point: CH_3O-H (100), CH_3-OH (~90), $HOCH_2-H$ (~90) $HO-H$ (116), C_6H_5-H (102), $C_6H_{11}-H$ (~94) (JOSCHEK and GROSSWEINER 1966, JOSCHEK and MILLER 1966).

There are marked differences in the ease of loss of halogen from an aromatic system. The orientation and nature of the other ring substituents have a profound effect on the course of the photochemical reaction. The solvent and the wavelength of the source of illumination are important in determining the nature of the products. A number of solvents such as cyclohexane, primary and secondary alcohols, and ether act as hydrogen donors and, in the photolytic reaction, halogen is replaced by hydrogen. Photolysis in water usually affords derivatives in which halogen has been replaced by hydroxyl but hydrogen abstraction is also observed. Benzene is not a good hydrogen donor: on photolysis a new C–C bond is normally formed and the departing halogen is replaced by a phenyl group. A variety of side reactions may also occur and sequential reactions of products often afford complex reaction mixtures. Examples of several of these influences have been investigated in a number of model systems.

The isomeric chlorobenzoic acids were irradiated in methanol by a 450-watt mercury arc with a Corex filter. After eight hours o-chlorobenzoic acid gave an almost quantitative yield of benzoic acid, p-chlorobenzoic acid gave 30 percent yield of benzoic acid and 70 percent of unchanged starting material, m-chlorobenzoic acid gave only 5 percent of benzoic acid and 95 percent of starting material. Different results were obtained in isopropanol: o-chlorobenzoic acid gave a 50 percent yield of benzoic acid in eight hours, but p-chlorobenzoic acid gave an 80 percent yield; m-chlorobenzoic gave 10 percent benzoic acid. In this solvent side reactions consumed the remainder of the starting material. o-Chlorobenzoic acid gave a 25 percent yield of benzoic acid after eight hours irradiation in tertiary butanol. m- and p-Chlorobenzoic acid were recovered unchanged from this solvent with approximately one percent conversion to benzoic acid (PLIMMER and HUMMER 1968).

Irradiation of the sodium salt of o-chlorobenzoic acid in water gave unchanged starting material with one to two percent of benzoic acid and one to two percent salicylic acid after eight hours. The photochemical reaction is much slower in water and both hydroxylation and hydrogen abstraction occur. The energies required for fission of O–H bonds in water and C–H bonds in methanol are 116 and 90 kcals., respectively. The reaction which requires least energy is the breaking of the C–H bond in methanol. Confirmation of this hypothesis is afforded by studies of the reaction in isopropanol, tertiary butanol, diethyl ether, and ethanol. Solvents containing the H–C–O– group most readily act as hydrogen donors, whereas the reaction appears blocked in tertiary butanol.

The behavior of the herbicide, 2,6-dichlorobenzonitrile (dichlobenil) (5), has been studied in some detail to determine the influence of the cyano group on the course of the reaction (PLIMMER and HUMMER 1968 b). Dichlobenil was irradiated in methanol with a 450-watt mercury lamp and a Corex filter. The composition of the reaction mixture was analyzed at intervals by gas chromatography. Although the cyanide radical may behave as a pseudo-halogen, the only products obtained were o-chlorobenzonitrile (6) and benzonitrile (7). There was

(5)   (6)   (7)

no loss of CN group. After eight hours the major product of photolysis was o-chlorobenzonitrile. Some benzonitrile was also obtained. The conversion of chlorobenzonitrile to benzonitrile was much slower than the conversion of dichlobenil to o-chlorobenzonitrile. As the products of photochemical processes may be transformed in subsequent reactions, the composition of the reaction mixture at any given time will be dependent on the rate constants of these reactions.

The relationship between the absorption process and the possibility of photochemical reaction was mentioned in the introductory section. It is of interest to compare the absorption coefficient and wavelength maxima in methanol for dichlobenil, o-chlorobenzonitrile, and benzonitrile (Table I). Dichlobenil was wavelength maxima at 2,980 and

Table I. *Wave length maxima and absorption coefficients* (in methanol)

Compound	λ Max. (A)	ε Max.
Dichlobenil	2,980	2,100
	2,880	2,000
o-Chlorobenzonitrile	2,880	1,380
	2,800	1,380
Benzonitrile	2,770	840
	2,700	900
	2,630	620

2,880 A whereas o-chlorobenzonitrile has shorter wavelength maxima at 2,880 and 2,800 A of considerably lower extinction coefficients. The former compound is a more efficient absorber of light above 2,800 A. Studies of quantum yields of a number of photochemical reactions of this type would be valuable as an aid to understanding the photochemical processes.

The influence of aromatic substituents in the ring on loss of halogen was investigated by SZYCHLINSKI and his co-workers (SZYCHLINSKI 1963, SZYCHLINSKI and LITWIN 1963, WALCZAK and SZYCHLINSKI 1963). The mechanism and kinetics of a number of photolyses of halogenated benzene derivatives were the subject of a physico-chemical study. The isomeric chlorobenzoic acids were irradiated in methanol, and the liberated chloride ion was determined by titration. o-Chlorobenzoic acid was most reactive. Irradiation of bromobenzene in methanol gave only a trace of benzene. The effect of viscosity and electrolyte concentration on the reaction was determined. No exchange was observed to occur between isotopically labeled bromine atoms (present in the ionic form) and the bromine atoms originally present in the bromobenzene.

The relative reactivity of the halogen atoms in the chlorobenzoic acids and dichlobenil provided a starting point for investigation of the relative ease of loss of chlorine from a number of halogenated aromatic compounds. Preliminary results of these studies have been reported (PLIMMER and HUMMER 1968 a and b).

The herbicide 2,3,6-trichlorobenzoic acid (8), is an example of a polychlorinated compound. It was predicted that this compound would react in methanol by loss of a single chlorine atom to give a mixture of three possible dichlorobenzoic acids.

Subsequent reaction of the dichlorobenzoic acids would afford o-
and m-chlorobenzoic acids: ultimately benzoic acid should result. The
amounts of individual photoproducts would depend on the relative
rates of decomposition of the intermediates. The 2,6- and 2,5- di-sub-
stituted dichlorobenzoic acids were first studied. 2,6-Dichlorobenzoic
acid (9) lost a chlorine atom to give o-chlorobenzoic acid but the loss
of chlorine from o-chlorobenzoic acid to give benzoic acid was much
more rapid. 2,5-Dichlorobenzoic acid (11) afforded almost exclusively
m-chlorobenzoic acid. The photolysis of 2,3,6-trichlorobenzoic acid
gave only 2,6- and 2,5-dichlorobenzoic acids as the only detectable
dichlorobenzoic acids. If any 2,3-dichlorobenzoic acid (10) was formed
it broke down very rapidly to m-chlorobenzoic acid. In the late stages
of the photolyses the quantities of benzoic acid and m-chlorobenzoic
acid present in the reaction mixture increased as would be predicted.

To extend these observations, the influence of other substituents on
loss of chlorine was determined by measuring yields of photoproduct
and decrease in quantity of starting material. For ortho-substituents
the order of effects was $CO_2H > OCH_3 > CH_3$ (ease of loss of chlorine
decreasing); for para-substituents $OCH_3 \gg CO_2H = CH_3$, and for
meta-substituents $CH_3 > OCH_3 \gg CO_2H$ (PLIMMER and HUMMER
1968 b).

Rationalization of photochemical data in terms of known electronic
effects has been reported in one study (ZIFFER and SHARPLES 1962) in
which the rate of photochemical loss of nitrogen by substituted aromatic diazoketones was correlated with the Hammett function for a
given substituent. The reaction is known to proceed by a triplet excited state in which the electronic distribution is thought to approximate that of the ground state and application of the Hammett function
may be valid in this case.

The excited state in halogen loss is probably a singlet and little information exists concerning electron distribution in this state. Detailed
investigations of the photochemistry and photochemical reactivity of
chlorinated aromatic compounds are required to enhance the predictive
value of empirical studies.

IV. The role of oxygen

It is safe to assume that in the majority of irradiation experiments
described in this review that oxygen was present in solution. The dif-
ficulties in removing oxygen completely have been discussed (CALVERT
and PITTS 1966) and elaborate techniques are required. The oxygena-
tion of products may occur directly or by a sensitized mechanism. The
mechanism of dye-sensitized photooxidation has been recently reviewed and it is suggested that singlet oxygen may be involved in this
process (FOOTE 1968 a and b). Further studies of the effects of sensitizers and investigations of highly purified systems are called for be-

fore the practical significance of trace impurities and sensitized reactions can be assessed.

The effect of oxygen on the photochemical reaction of iodobiphenyl in benzene has been studied in some detail and, if the solution was kept saturated with oxygen during the photolyses, a 26 percent yield of 4-hydroxybiphenyl was obtained (KHARASCH and SHARMA 1966). Photolysis of iodobiphenyl in benzene gave terphenyl by reaction of the photolytically generated radical with the solvent or biphenyl by hydrogen abstraction. The product ratio (biphenyl:terphenyl) was highest for a rigorously degassed solution. Therefore, in the presence of oxygen there may be a decrease in the amount of hydrogen-transfer product, as oxygen could react competitively with the biphenyl radical to give 4-hydroxybiphenyl by way of the intermediate hydroperoxide.

The effects of degassing fluoro-, chloro-, and bromobiphenyls differed. A high yield of biphenyl could be obtained from 4-bromobiphenyl without rigorous degassing (KHARASCH et al. 1966). Possibly oxygen may have a significant role in some of the photohydroxylation reactions of herbicides. Hydroperoxides are potential intermediates and it is of interest to establish whether they are involved in this reaction.

V. Specific Herbicides

a) 3-Amino-2,5-dichlorobenzoic acid (amiben)

Amiben (12) is readily susceptible to photolysis by visible light. Sunlight or a fluorescent sunlamp caused an aqueous solution of amiben to become colored. A number of products could be separated by thin-layer or paper chromatography and some of them were phenolic (SHEETS 1963, CROSBY 1966). Amiben and its methyl ester on prolonged exposure to light gave a deep colored solution. A number of photolytic processes could occur in a complex molecule of this type; it was clear from examination of the products by chromatography and spectroscopy that polymerization had taken place to give colored products of high molecular weight (PLIMMER and HUMMER 1967). Condensation can occur in several ways. For example, replacement of the chlorine atom at the 2-position by hydroxyl would afford a 2-aminophenol, and condensation of two molecules of 2-aminophenol to a phenoxazine would follow. Loss of chlorine gives free radicals which may form di- or polyphenyl derivatives. The oxidation of the amino-group by loss of a hydrogen atom would give a radical which could dimerize with a second molecule to give azo compounds via a head-to-head union or polymers via a nuclear attack. Such processes are brought about in aromatic amines photochemically (SANTHANAM and RAMAKRISHNAN 1968) or by free-radical reactions (DANIELS and SAUNDERS 1953).

Many of these reactions represent secondary reactions of the initial photoproducts. Inhibition of oxidation by sulfur dioxide (as bisulphite), during photolysis of amiben in water or amiben methyl ester in methanol, reduced secondary reactions. Illumination by a lamp with a borosilicate glass filter rapidly dechlorinated the compounds at the 2-position. A comparatively high yield of methyl 3-amino-5-chloroben-zoate (82 percent) was obtained from the ester in one hour (PLIMMER and HUMMER 1969). The free acid in water also underwent reductive dechlorination. Colored substances accompanied the major product indicating that hydroxylation had occurred. A trace of the completely dechlorinated 3-aminobenzoic acid (14) was obtained but 31 percent of the 3-amino-5-chlorobenzoic acid (13) was obtained. It is note-

worthy that under these conditions abstraction of hydrogen from water was preferred to hydroxylation.

The biological activity of amiben diminishes during irradiation. Modification of the primary amino group by formation of the N-benzoyl derivative gave a herbicidally active compound which was more stable toward light. Biological activity was retained and the measurement of the ultraviolet absorption spectrum indicated that less profound changes had occurred than in the case of amiben irradiated under similar conditions (ISENSEE et al. 1969).

b) 4-Amino-3,5,6-trichloropicolinic acid (picloram)

Picloram (15) is degraded by light under a variety of conditions. Breakdown by light takes place in solution and in the solid phase (REDEMAN 1968). After 48 hours 60 percent of one mg. of picloram applied to the surface of a petri dish had been degraded by ultraviolet light. Under the same conditions a 35 percent breakdown resulted in sunlight. After one week these figures had increased to 90 and 65 percent, respectively. Degradation on a soil surface was slower (MERKLE et al. 1967).

Breakdown occurred rapidly in aqueous solution on exposure to ultraviolet light (KEARNEY et al. 1968). Under irradiation at 2,537 A, two molecules of chloride ion were liberated for each molecule of picloram degraded and acidic compounds were formed. At least five methylated derivatives of the decomposition products were separated by chromatography (HALL et al. 1968).

While the products of picloram photolysis in water have not been isolated, it seems reasonable to predict that replacement of chlorine by hydroxyl takes place. It is probable that initial reaction occurs at the 6-position. 4-Amino-3,5-dichloro-6-hydroxypicolinic acid (16) has recently been isolated as a metabolite of picloram in wheat (REDEMANN et al. 1968), and this position would be expected to be more reactive than the 3- and 5- positions.

Under ultraviolet irradiation and in the presence of oxygen, pyridine is photolysed in water to a glutaconic aldehyde derivative, suggesting that oxidative breakdown may be a possible route for more complex derivatives (JOUSSOT-DUBIEN and HOWARD 1967).

An interesting parallel is the insecticide O,O-diethyl O-3,5,6-trichloro-2-pyridyl phosphorothioate (17) (Dursban [3]) which is a derivative of 3,5,6-trichloro-2-pyridinol (18). The photolysis of 3,5,6-trichloro-2-pyridinol in sunlight and ultraviolet light with a sunlamp was shown to yield dehalogenated products (SMITH 1968). Liberation of chlorine was followed by using ^{36}Cl-3,5,6-trichloropyridinol and subjecting the reaction mixture to radiochemical analysis. In the dry state little decomposition occurred. In buffered aqueous solution (pH 8.0) 3,5,6-trichloropyridinol was rapidly degraded. All the chlorine atoms appeared to be liberated at about the same rate but it was suggested that the 6-chlorine atom may be liberated first. Measurement of absorbancy indicated that loss of halogen and ring cleavage did not occur

---

[3] Registered trademark of Dow Chemical Co.

simultaneously. Ring cleavage followed and it was postulated that diols
and triols could be formed. Fourteen products were detected, but isola-
tion of these was made difficult by their readiness to oxidize. The end
products of oxidation were carbon dioxide and a series of colored products as in the scheme proposed by the investigators (see above).

The photolysis of methyl 4-amino-3,5,6-trichloropicolinate in methanol irradiated by an ultraviolet lamp with a Corex glass filter gave a
mixture of products. Preliminary investigations have shown that at
least two monodehalogenated products are present. These compounds
were separated and identified by a combination of gas chromatography
and mass spectrometry. The decarboxylation product of picloram, 4-
amino-3,5,6-trichloropyridine (19), was also found to be present in the
reaction mixture (PLIMMER and HUMMER 1968 c).

c) Chlorinated benzoic acids

The photochemistry of the most important herbicidal chlorinated
benzoic acid, 2,3,6-trichlorobenzoic acid (2,3,6-TBA) (8), in methanol was discussed in a previous section. Other observations relating
to the chlorinated benzoic acids were discussed in the same section.
The photochemistry of the monochlorinated benzoic acids in water has
practical relevance and has been investigated by CROSBY (1966). Irradiated at 2,540 A the sodium salts in aqueous solution underwent
replacement of halogen by hydroxyl or hydrogen. Several other compounds were isolated and p-chlorobenzoic acid afforded, in addition
to p-hydroxybenzoic acid and benzoic acid, two compounds identified
as p-acetylbenzoic acid and terephthalic acid. The source of the additional carbon atoms is open to speculation. The only additional source
of carbon, in addition to the starting material, was carbon dioxide,
arising from initial neutralization of sodium carbonate by the free acid
to form the salt.

It is to be anticipated that the end products of photolysis of the
chlorinated benzoic acids will be mixtures of reduced and hydroxylated
acids. The photochemistry of 3-amino-2,5-dichlorobenzoic acid is discussed in a separate section. There is little information concerning
other chlorinated aromatic acids, such as 3,6-dichloro-2-methoxy-
benzoic acid (dicamba) (20) but in the course of a preliminary survey,
this compound appeared fairly resistant to photodecomposition (PLIM-
MER and HUMMER 1968 c).

(20)   (21)   (22)

d) Chlorinated phenylacetic acids

The herbicide fenac is a mixture of chlorinated phenylacetic acids. The commercial mixture contains predominantly trichlorophenylacetic acids, of which the 2,3,5-isomer is the most important (21). The photolysis of such a mixture gave, as would be expected, a complex mixture of products. In aqueous solution irradiation of the sodium salt of fenac by ultraviolet light afforded a complex mixture of acidic and neutral compounds. Several of these were tentatively identified. The major product was probably 2,5-dichlorobenzyl alcohol (22): the facile loss of the 2-chloro group from a 2,3,6-trichloro-substituted compounds affords an interesting parallel to the case of 2,3,6-trichlorobenzoic acid where the initial loss of the 2-chloro substituent also occurs. Di- and trichlorobenzaldehydes were probable minor products. Reduction and decarboxylation accompany loss of chlorine but some of the products are themselves photolabile. A study of the individual isomeric monochlorophenylacetic acids provided additional information relevant to the behavior of the more complex derivatives. For example, the irradiation of 2-chlorophenylacetic acid as the sodium salt in water gave benzyl alcohol, benzaldehyde, and o-chlorobenzaldehyde as neutral products. In addition, 2-methoxyphenylacetic acid, 2-methoxybenzyl alcohol, and possibly phenylacetic acid were identified by methylation of the acidic function. Photochemical transformations of the –CH$_2$CO$_2$H group occur simultaneously with reductive loss of chlorine and replacement by hydroxyl (CROSBY 1966).

e) 2,6-Dichlorobenzonitrile (dichlobenil)

Studies of 2,6-dichlorobenzonitrile photochemistry, discussed in the section on mechanism, suggest that this compound would be liable to decompose by loss of chlorine. The loss or replacement of one chlorine atom requires much less energy than the subsequent loss of a second chlorine atom.

f) 2,4-Dichlorophenoxyacetic acid (2,4-D)

The effect of sunlight on the behavior of 2,4-dichlorophenoxyacetic acid (2,4-D), (23) as a herbicide was studied by PENFOUND and MINYARD (1947). Later workers investigated the effect of ultraviolet light on the compound (MITCHELL 1961, ALY and FAUST 1964). The photosensitized decomposition of 2,4-D acid took place in the presence of riboflavin (BELL 1965). The major products of photolysis in aqueous solution have been isolated and identified. 2,4-Dichlorophenol (24) was isolated from the reaction mixture following ultraviolet irradiation and 4-chlorocatechol (25) could be identified by vapor-phase and thin-layer chromatography. The acidic fraction contained 4-chloro-2-

hydroxyphenoxyacetic acid (26) and a small quantity of 2-chloro-4-hydroxyphenoxyacetic acid (27). A number of acids was isolated but the photolysate did not appear to contain glycolic, succinic, or oxalic acids which might be expected to result by fission of the ether link. Residues of brown-colored materials were probably formed by oxidation of 1,2,4-trihydroxybenzene by a process independent of illumination. Addition of sodium bisulphite to the mixture inhibited this oxidation, and 1,2,4-trihydroxybenzene (28) was isolated and characterized as the acetate. Almost all the 2,4-D irradiated under laboratory conditions was eventually converted into brown polymeric material (or humic acid). This material was also produced in sunlight and the effect of sunlight was generally similar to that of ultraviolet irradiation. Chlorine atoms were replaced by hydroxyl groups and fission of the ether link occurred. The final step may be oxidation of 1,2,4-trihydroxybenzene to the intermediate 2-hydroxybenzoquinone (29) which then polymerizes (CROSBY and TUTASS 1966).

This study provides the model for other herbicides of this group and it is to be predicted that they will undergo photodecomposition to give similar products. 2-(2′,4′,5′-Trichlorophenoxy)propionic acid (Silvex), 4-(2′,4′-dichlorophenoxy)butyric acid (2,4-DB), 2,4,5-trichlorophenoxyacetic acid (2,4,5-T), and the isomeric monochlorophenoxyacetic acids afford the corresponding phenols on photolysis in aqueous solution (CROSBY and TUTASS 1965).

g) Ioxynil and bromoxynil

The herbicide, 3,5-diiodo-4-cyanophenol (ioxynil) (30) irradiated by ultraviolet light in benzene afforded 3,5-diphenyl-4-cyanophenol (31) (UGOCHUKWU and WAIN 1965). The introduction of phenyl

groups, typical of iodoaromatic compounds, has been discussed in an earlier section of this review. In aqueous solution 3,5-dibromo-4-cyano-phenol (bromoxynil) was decomposed by light to polyphenols and colored polymeric materials (CROSBY and TUTASS 1965).

h) Pentachlorophenol

The rice herbicide, pentachlorophenol (PCP) (32) as the sodium salt breaks down on exposure to sunlight. Ultraviolet irradiation in hexane or methanol gave 2,3,5,6-tetrachlorophenol (33) by reductive loss of chlorine whereas irradiation of a suspension of the free phenol in water afforded a little tetrachlorophenol (33), chloranil (34), and chloranilic acid (35). However, polymeric substances were the major photolysis products (HAMADMAD 1967). An aqueous solution was exposed to sunlight for ten days. From the violet colored solution a number of reaction products were isolated. The major products were chloranilic acid (35) and a yellow compound identified as 3,4,5-trichloro-6-(2'-hydroxy-3',4',5',6'-tetrachlorophenoxy)-o-benzoquinone (36). Further chromatography of the reaction mixture afforded quantities of minor products which were identified as tetrachlororesorcinol (0.10 percent) (37), 2,5-dichloro-3-hydroxy-6-pentachlorophenoxy-p-benzoquinone (38) (0.16 percent), and 3,5-dichloro-2-hydroxy-5-(2',4',5',6'-tetrachloro-3-hydroxyphenoxy-p-benzoquinone (39) (0.08 percent) (KUWAHARA et al. 1966 a and b).

The identification of products resulting from oxidative coupling of phenols and subsequent oxidation exemplifies the complexity of the reactions which may follow initial photochemical replacement of chlorine. Deeply colored photochemical products are frequently reported (cf. inter al. CROSBY and TUTASS 1966) and their production may be ascribed to similar oxidative coupling reactions.

i) s-Triazines

The photochemistry of the important s-triazine group of herbicides has not been thoroughly studied. A number of reports and reviews refer to their inactivation by light and also by infrared irradiation. It has not always been clear whether loss by volatilization might be responsible for reduced activity, rather than photodecomposition (ERCEGOVICH 1965).

Decomposition of 2-chloro-4-ethylamino-6-isopropylamino-s-triazine (atrazine) (40), 2-chloro-4,6-bis(ethylamino)-s-triazine (simazine) (41), and 2-methylmercapto-4-ethylamino-6-isopropylamino-s-triazine (ametryne) (42) occurred when the herbicides were adsorbed on filter paper and irradiated by ultraviolet light and sunlight. Measurement of the ultraviolet spectrum showed progressive changes dependent on the wavelength used (JORDAN et al. 1964 a).

Loss of halogen is to be expected by analogy with the behavior of chlorinated benzoic compounds. The triazines are, however, comparatively stable to light and require quite low wavelengths to disrupt the molecule. Recent investigations confirm that radiation of short wavelength is necessary to affect the dialkylamino-s-triazines (PLIMMER and KLINGEBIEL 1968). Light of wavelength 2,200 A was necessary to bring about photochemical reaction. Simazine and 4,6-bis(ethylamino)-2-methoxy-s-triazine (simetone) (43) were irradiated with ultraviolet light in methanolic solution. Simazine gave a number of products. The chloro group was displaced as would be predicted and 4,6-bis(ethylamino)-s-triazine (44) was obtained together with simetone (43). Products were separated by gas chromatography and evidence for structure was based on mass spectrometric studies. Following the replacement of chlorine by methoxyl the complete loss of the alkylamino side chain took place and this was replaced by hydrogen affording 4-ethylamino-2-methoxy-s-triazine (45). Some methylated products appear to be formed after loss of chlorine: a ring nitrogen atom may be methylated. Simetone followed a similar pattern. Loss of the side chain to give 4-ethylamino-2-methoxy-s-triazine (45) was an important reaction: methylation also occurred. The rate of photolysis of simetone was much slower than that of simazine and unchanged starting material was present after 12 hours.

j) Urea herbicides

The urea herbicides are derived from aniline. The aniline residue commonly bears one or more halogen atom substituents. During the investigation of the fate of urea herbicides in soils it was realized that photodecomposition might play some part in herbicide loss from the soil surface in dry areas (HILL et al. 1955). Loss of 83 percent of the initial quantity of 3-(p-chlorophenyl)-1,1-dimethylurea (monuron) (46) occurred during 48 days exposure of an aqueous solution, sealed in quartz, to sunlight (HILL et al. 1955). Ultraviolet light was shown to degrade 3-(3,4-dichlorophenyl)-1,1-dimethylurea (diuron) (47) and monuron (46) (WELDON and TIMMONS 1961). Diuron, monuron, 1-n-butyl-3-(3,4-dichlorophenyl)-1-methylurea (neburon) (48), and 3-phenyl-1,1-dimethylurea (fenuron) (49) were absorbed on filter paper and exposed to sunlight and ultraviolet light of wavelength 2,537 A. Changes in the absorption spectra indicated that decomposition had occurred (JORDAN et al. 1964 b). Greatest changes occurred with shorter wavelengths (JORDAN et al. 1965). The exposure of solid 2-^{14}C-labeled 3-(p-chlorophenoxy)-phenyl-1,1-dimethylurea (chloroxuron) (50) to a source of ultraviolet light resulted in 90 percent loss in 13 hours. Mono-(2.2 percent) (51) and di-demethylated (4.2 percent) (52) products were identified in the reaction mixture and 64 percent of the original radioactive label was evolved as carbon dioxide (GEISS-BÜHLER et al. 1963).

More detailed information concerning the identity of the products has been reported recently. The replacement of halogen by hydroxyl has been observed in a number of examples. The products of irradiation of an aqueous solution of 3-(p-bromophenyl)-1-methoxy-1-methyl-urea (metobromuron) (53) exposed to sunlight for 17 days were isolated and identified. Starting material accounted for 80 percent of the mixture and 15 percent was 3-(p-hydroxyphenyl)-1-methoxy-1-methyl-urea. The remaining material contained 3-(p-bromophenyl)-1-methyl-urea, p-bromophenylurea, and some unidentified products. Qualitatively similar results were obtained by irradiating an aqueous solution with a low-pressure mercury lamp (ROSEN and STRUSZ 1968).

Aqueous solutions of monuron and 3-(3,4-dichlorophenyl)-1-methoxy-1-methylurea (linuron) (54) were exposed to sunlight for varying periods. Linuron gave (after two months) 13 percent 3-(3-chloro-4-hydroxyphenyl)-1-methoxy-1-methylurea (55), 10 percent 3,4-dichlorophenylurea (56), and two percent 3-(3,4-dichlorophenyl)-1-methylurea (57). The chlorine substituent *para* to the amino group was replaced by hydroxyl in this case (ROSEN et al. 1969).

One of the products of monuron photolysis was 3-(p-hydroxy-phenyl)-1,1-dimethylurea (58), as would be predicted from the normal course of the reaction. However, in another laboratory the photolysis

NHCON(CH₃)(OCH₃) — $NHCON(CH_3)(OCH_3)$

$NHCON(CH_3)(OCH_3)$

Br

(53)

$NHCON(CH_3)(OCH_3)$

Cl

Cl

(54)

$NHCON(CH_3)(OCH_3)$

Cl

OH

(55)

$NHCONH_2$

Cl

Cl

(56)

$NHCONHCH_3$

Cl

Cl

(57)

$NHCON(CH_3)_2$

OH

(58)

$NHCON(CH_3)_2$

OH

Cl

(59)

$NHCONH$

Cl— —Cl

(60)

of monuron under similar conditions gave rise to different products. A number of oxidized and polymerized compounds were obtained but 3-(4-chloro-2-hydroxyphenyl)-1,1-dimethylurea (59) was the only hydroxylated product. It is conceivable that this product might arise by rearrangement following an initial oxidation of the nitrogen atom. The reversal of the synthetic reaction by which monuron may be formed was also reported to occur during the photolysis. The elimination of dimethylamine results in formation of 4-chlorophenyl isocyanate; further reaction of this compound with 3,4-dichloroaniline, also produced during photolysis, afforded 1,3-di(p-chlorophenyl)urea (60) which was isolated from the reaction mixture (TANG and CROSBY 1968). This parallels the result of photolysis of isopropyl-N-phenylcarbamate (IPC) which gave sym-diphenylurea by a similar mechanism (CROSBY 1966). The difference in the findings of the two laboratories emphasizes the necessity for inclusion of all relevant experimental details in reports of photochemical reactions.

k) Miscellaneous

The photochemistry of a number of important halogenated herbicides has not been reported. Some of these are currently under investigation. In many other cases only the most superficial information exists as literature reports deal with inactivation and decomposition by light and are primarily of interest to the plant physiologist. From the information previously summarized the behavior of several of these herbicides can be inferred to a limited extent, however.

The amides, 3'4'-dichloropropionanilide (propanil), 3'-chloro-2-methyl-p-valerotoluidide (solan), and a number of related compounds are widely used in quantity, but there is little recorded information concerning their photochemical behavior. The loss of chlorine from these compounds is to be predicted and amongst other pathways the photoanilide rearrangement, in which the acyl group migrates from the nitrogen atom to the adjacent o-carbon atom of the benzene ring, seems likely (ELAD et al. 1965).

The herbicidal carbamate, isopropyl-m-chlorocarbanilate (CIPC, chlorpropham) can lose chlorine or alternatively yield p-chlorophenyl isocyanate by loss of isopropanol. Further reactions might yield products analogous to those likely to arise from IPC.

Solid films of 5-bromo-3-sec-butyl-6-methyluracil (bromacil) are reported to decompose in ultraviolet light (JORDAN et al. 1965), and a solution of this herbicide in water is slowly inactivated by ultraviolet light (KEARNEY et al. 1968). The chemistry of the reaction has not been further studied but this and many similar reports provide material for further investigation.

A number of herbicidal compounds carry trifluoromethyl substituents: the reviewer has not attempted to include any of these in this survey and reports of their photochemistry are few.

VI. Conclusion

Model studies of a variety of halogenated aromatic compounds have demonstrated that photodecomposition will occur if light of sufficient energy is available. Light from a mercury arc (2,540 A) is adequate in the majority of cases. Loss of halogen is the dominant process. Studies of the influence of other substituents on the aromatic ring are in an early stage, but ease of loss of halogen is profoundly affected by the orientation and electronic effect of a substituent. Photoreaction generates a halogen atom and a free phenyl radical. The reactions of the free phenyl radical depend on the environment. In dilute aqueous solution products are formed by hydroxylation or replacement of chlorine by hydrogen. Hydrogen abstraction is favored in donor solvents such as methanol and cyclohexane. In benzene, phenylation will occur.

It is much more difficult to assess the effects of photodecomposition on performance of a herbicide in the field or the practical importance of photochemical transformation products. It is noteworthy, however, that in at least one case the products isolated from the plant, following the application of a herbicide, were identical with the products of photochemical decomposition (SLADE 1968). A major problem in relating loss of herbicidal activity in the field to photodecomposition is the number of competing modes of loss, such as volatilization, microbial breakdown, and leaching. Loss by volatilization, except in specific cases, is very difficult to assess even under carefully controlled laboratory conditions.

Examination of the absorption spectrum of the solution and consideration of the chemical structure of a herbicide is generally sufficient to indicate whether photodecomposition is likely to occur. Photodecomposition is not a necessary consequence of light absorption and many other pathways of degradation of light energy may be encountered. There is continuously increasing interest in the investigation of photochemical reactions from both the organic and physical chemical standpoints.

Although photodecomposition of a herbicide may not be of practical importance as a factor in its inactivation or disappearance, photochemical processes are nevertheless significant as a feature of overall chemical reactivity. The production of a radical by a photolytic or other homolytic process may be an important factor in the biological action of a halogenated herbicide. For example, it has been suggested that 3,5-diiodo-4-hydroxybenzonitrile may function herbicidally by acting as a decoupler in the photosynthetic electron transfer process (UGOCHUKWU and WAIN 1965). Whether this hypothesis is correct or not, there is need for more fundamental information. The quantum yield of the reaction and the relative bond strengths of carbon-halogen bonds in polysubstituted aromatic compounds may provide more relevant factual information.

Bond fission may result from direct absorption of light energy or the energy for bond fission may be supplied by a sensitizer. What compounds may function as energy transfer agents in this process? There is a need for assessing potential sensitizers in the environment and studying their part in transformation of herbicides. What is the physical state of the herbicide on the foliar surface and what is the nature of its chemical environment? Factors such as these will determine the nature of photoproducts. What is the effect of physical state on light absorption and photodecomposition?

The answers to these questions are necessary for an understanding of herbicide chemistry in the context of environment studies. The scientist is concerned with the activity and fate of an applied herbicide: it may remain in the soil, on the plant, or be volatilized into the atmosphere. It may be transformed biologically or chemically. Photochemical conversions are part of the overall scheme and adequate knowledge of the chemistry and toxicology of photochemical products is essential for the assessment of the performance and value of a herbicide.

Studies have been carried out frequently on model systems. The information they afford has often been of limited usefulness in relation to a specific field situation, but it is of great value from the chemical or predictive standpoint. An increasing number of studies indicate that valuable parallels exist between biochemical and photochemical processes. Frequently, both involve homolytic reactions and the course of metabolism may be predictable from the reactivity of bonds, as deduced from photochemical studies.

It is becoming increasingly clear that a broader fundamental knowledge of the chemistry of herbicide molecules must accompany studies of their mode of action and provide the foundation for investigations of their fate in the environment. Photochemical studies are important in this context. Halogenated aromatic herbicides present varied types of molecules. Although considered purely as chemical types these are rarely novel, photochemical research in this field is of relatively recent standing, and many facts still remain to be ascertained.

Table II. *Common names of herbicides mentioned in text* (Weed Science Society of America)

Common name	Chemical name
ametryne	2-(ethylamino)-4-(isopropylamino)-6-(methylthio)-s-triazine
amiben	3-amino-2,5-dichlorobenzoic acid
atrazine	2-chloro-4-(ethylamino)-6-(isopropylamino)-s-triazine
bromacil	5-bromo-3-sec-butyl-6-methyluracil
bromoxynil	3,5-dibromo-4-hydroxybenzonitrile
chloroxuron	3-[p-(p-chlorophenoxy)phenyl]-1,1-dimethylurea
chlorpropham (CIPC)	isopropyl m-chlorocarbanilate
dicamba	3,6-dichloro-2-methoxybenzoic acid
dichlobenil	2,6-dichlorobenzonitrile
diuron	3-(3,4-dichlorophenyl)-1,1-dimethylurea
fenac	2,3,6-trichlorophenylacetic acid
fenuron	1,1-dimethyl-3-phenylurea
ioxynil	4-hydroxy-3,5-diiodobenzonitrile
IPC	isopropyl-N-phenylcarbamate
linuron	3-(3,4-dichlorophenyl)-1-methoxy-1-methylurea
metobromuron	3-(p-bromophenyl)-1-methoxy-1-methylurea
monuron	3-(p-chlorophenyl)-1,1-dimethylurea
neburon	1-butyl-3-(3,4-dichlorophenyl)-1-methylurea
PCP	pentachlorophenol
picloram	4-amino-3,5,6-trichloropicolinic acid
propanil	3',4'-dichloropropionanilide
silvex	2-(2,4,5-trichlorophenoxy)propionic acid
simazine	2-chloro-4,6-bis(ethylamino)-s-triazine
simetone	2,4-bis(ethylamino)-6-methoxy-s-triazine
solan	3'-chloro-2-methyl-p-valerotoluidide
2,3,6-TBA	2,3,6-trichlorobenzoic acid
2,4-D	2,4-dichlorophenoxyacetic acid
2,4,5-T	2,4,5-trichlorophenoxyacetic acid

Summary

Current knowledge concerning the photochemistry of the halogenated herbicides is summarized. The absorption of light by halogenated aromatic compounds is discussed. The energy absorbed is sufficient to break the C–X bond and free radicals are produced. The radicals react with the solvent and, in methanol, hydrogen abstraction occurs. In benzene a phenylated derivative results. Water affords hydroxylated and reduced products. Mechanistic studies are reviewed. The ease of loss of halogen depends on ring substituents and some information on their influence is available. The course of a number of photochemical reductions affords information of predictive value. Oxygen may play a role in the reaction and determine the nature of the products.

Reports of investigations of specific herbicides are summarized. Amiben, picloram, chlorinated benzoic acids, chlorinated phenylacetic acids, dichlobenil, 2,4-D, ioxynil and bromoxynil, pentachlorophenol, triazines, and ureas are discussed in detail. The relationship between behavior in the field and the results of model experiments in the laboratory is difficult to assess, but the latter studies are a necessary part of the overall study of herbicide chemistry. The findings are very relevant to our knowledge of metabolism, toxicology, and mode of action of herbicides.

Résumé *

Photochimie des desherbants halogénés

Les connaissances courantes concernant la photochimie des desherbants halogénés sont résumées. Il est discuté de l'absorption de la lumière par les composés aromatiques halogénés. L'énergie absorbée est suffisante pour rompre les liaisons C–X et former ainsi des radicaux libres. Les radicaux réagissent avec le solvant et un arrachement d'hydrogène se produit dans le méthanol. Dans le benzène il en résulte un dérivé phénylé. L'eau produit des produits hydroxylés et réduits. Les études de mécanisme sont décrites. La facilité de la perte d'halogène dépend des substituants du noyau et nous disposons de quelques informations sur leur influence. Le mécanisme d'un certain nombre de réducteurs photochimiques donne des informations utilisables aux prévisions d'autres phénomènes. L'oxygène peut jouer un rôle dans la réaction et déterminer la nature des produits.

Les résultats de travaux sur des desherbants spécifiques sont résumés. Amiben, picloram, acides chlorobenzoïques, acides chlorophényl acétiques, dichlobenique, 2,4-D, ioxynil et bromoxynil, pentachlorophénol, triazines et urées sont examinés en détail. La relation entre le comportement sur le terrain et les résultats des essais au laboratoire est difficile à fixer, mais ces dernières études font nécessairement partie de la chimie d'ensemble des desherbants. Les résultats acquis viennent compléter nos connaissances du métabolisme, de la toxicologie et du mode d'action des desherbants.

---

* Traduit par R. MESTRES.

Zusammenfassung *

Photochemie der halogenierten Unkrautvertilgungsmittel

Gegenwärtige Kenntnisse, die die Photochemie der halogenierten Unkrautvertilgungsmittel betreffen, werden zusammengefasst. Die Adsorption des Lichtes durch halogenierte aromatische Verbindungen wird diskutiert. Die adsorbierte Energie reicht aus, die C–X-Bindung zu brechen, und freie Radikale werden produziert. Die Radikalen reagieren mit dem Lösungsmittel, und in Methanol tritt Wasserstoff-Abstraktion auf. In Benzol entsteht ein phenyliertes Derivat. Wasser liefert hydroxylierte und reduzierte Produkte. Mechanistische Studien werden revidiert. Die Leichtigkeit des Halogenverlustes hängt von den Ringsubstituenten ab, und einige Information über ihren Einfluss ist vorhanden. Der Gang einer Anzahl photochemischer Reduktionen liefert Informationen von vorhersagendem Wert. Sauerstoff kann in der Reaktion eine Rolle spielen und die Natur der Produkte bestimmen.

Berichte über Untersuchungen spezifischer Unkrautvertilgungsmittel werden zusammengefasst. "Amiben, picloram, chlorinated benzoic acids, chlorinated phenylacetic acids, dichlobenil, 2,4-D, ioxynil und bromoxynil, pentachlorophenol", Triazine und Harnstoffe werden im einzelnen diskutiert. Eine Beziehung zwischen dem Verhalten im Feld und den Resultaten von Modell-Versuchen im Laboratorium ist schwierig abzuschätzen, aber die letzteren Forschungen sind ein notwendiger Teil des gesamten Studiums der Unkrautvertilgungsmittel-Chemie. Die Entdeckungen sind sehr wesentlich für unsere Kenntnisse des Stoffwechsels, der Toxikologie und der Wirkungsweise der Unkrautvertilgungsmittel.

References

ALY, O. M., and S. D. FAUST: Studies on the fate of 2,4-D and ester derivatives in natural surface waters. J. Agr. Food Chem. **12**, 541 (1964).

BARLTROP, J. A., N. J. BUNCE, and A. THOMSON: Organic photochemistry. Part IV. Photonucleophilic substitution: Reactions of monosubstituted benzenes. J. Chem. Soc. C, p. 1142 (1967).

BELL, G. R.: Photochemical degradation of 2,4-dichlorophenoxyacetic acid and structurally related compounds in the presence of riboflavin. Botan. Gaz. **118**, 133 (1956).

BLAIR, J. McDONALD and D. BRYCE-SMITH: Liquid phase photolysis. Part II. Iodobenzene. J. Chem. Soc., p. 1788 (1960).

—— ——, and B. W. PENGILLY: Liquid phase photolysis. Part I. Variation of isomer ratios with radical source in the phenylation of isopropylbenzene. Photolytic generation of phenyl radicals. J. Chem Soc., p. 3174 (1959).

CALVERT, J., and J. N. PITTS, JR.: Photochemistry. New York: Wiley (1966).

COMES, R. D., and F. L. TIMMONS: Effect of sunlight on the phytotoxicity of some phenylurea and triazine herbicides on a soil surface. Weeds **13**, 81 (1965).

---

* Übersetzt von M. DÜSCH.

CROSBY, D. G.: Photochemistry of herbicides. 154th Amer. Chem. Soc. Nat. Meeting, New York (1966).
——, and M-Y LI: Herbicide photodecomposition. In: Degradation of herbicides. P. C. Kearney and D. D. Kaufman (eds.). New York: Dekker (1969).
——, and H. O. TUTASS: Photodecomposition of 2,4-dichlorophenoxyacetic acid. J. Agr. Food Chem. 14, 596 (1966).
DANIELS, D. G. H., and B. C. SAUNDERS: Studies in peroxidase action. Part VIII. The oxidation of p-chloroaniline. A reaction involving dechlorination. J. Chem. Soc., p. 822 (1953).
ELAD, D., D. V. RAO, and V. I. STENBERG: The photoanilide rearrangement. J. Org. Chem. 30, 3252 (1965).
ERCEGOVICH, C. D.: What happens to the triazines in soil? Ardsley, New York, p. 22 (1965).
EULER, H.: Über die Lichtspaltung von Halogen-Essigsäuren in Benzol und Äther. Chem. Ber. 49, 1366 (1916).
FOOTE, C. S.: Photosensitized oxygenations and the role of singlet oxygen. Accounts Chem. Research 1, 104 (1968 a).
—— Mechanisms of photosensitized oxidation. Science 162, 963 (1968).
GEISSBÜHLER, H., HASELBACH, H. AEBI, and L. EBNER: The fate of N'-(4-chlorophenoxy)-phenyl-NN-dimethylurea (C-1983) in soils and plants. III. Breakdown in soils and plants. Weed Research 3, 277 (1963).
HALL, R. C., C. S. GIAM, and M. G. MERKLE: The degradation of picloram by light. Weed Sci. Soc. Amer. Nat. Meeting, New Orleans, La., Abstr. p. 20 (1968).
HAMADMAD, N.: Photolysis of pentachloronitrobenzene, 2,3,5,6-tetrachloronitrobenzene, and pentachlorophenol. Thesis, Univ. of Calif., Davis, Calif. (1967) [Quoted in CROSBY (1969), ref. 67].
HENDERSON, W. A., JR., and G. ZWEIG: Photolytic rearrangement and halogen dependent photocyclization of halophenyl naphthalenes. J. Amer. Chem. Soc. 89, 6778 (1967).
ISENSEE, A. R., J. R. PLIMMER, and B. C. TURNER: The effect of light on the herbicidal activity of some amiben derivatives. Weed Sci. Soc. Amer. Nat. Meeting, Las Vegas, Nev., Abstr. No. 224 (1969)
JORDAN, L. S., B. E. DAY, and W. A. CLERX: Photodecomposition of triazines. Weeds 12, 5 (1964 a).
——, C. W. COGGINS, B. E. DAY, and W. A. CLERX: Photodecomposition of substituted phenylureas. Weeds 12, 1 (1964 b).
——, J. D. MANN, and B. E. DAY: Effects of ultraviolet light on herbicides. Weeds 13, 43 (1965).
JOSCHEK, H. I., and L. I. GROSSWEINER: Optical generation of hydrated electrons from aromatic compounds. II. J. Amer. Chem. Soc. 88, 3261 (1966).
—— —— Photocleavage of phenoxyphenols and bromophenols. J. Amer. Chem. Soc. 88, 3269 (1966).
JOUSSOT-DUBIEN, J., and J. HOUDARD: Reversible photolysis of pyridine in aqueous solution. Tetrahedron Letters, p. 4389 (1967).
KEARNEY, P. C., E. A. WOOLSON, J. R. PLIMMER, and A. R. ISENSEE: Decontamination of pesticide residues in soils. Residue Reviews 29, 137 (1969).
KHARASCH, N., and R. K. SHARMA: Oxygen effects in the photolysis of 4-iodobiphenyl in benzene. Chem. Commun., p. 106 (1966).
—— ——, and H. B. LEWIS: The photolysis of 4-bromobiphenyl in benzene. Chem. Commun., p. 418 (1966).
KOLLER, L. R.: Ultraviolet radiation (2ed.). New York: Wiley (1966).
KUWAHARA, M., N. KATO, and K. MUNAKATA: The photochemical reaction of pentachlorophenol. Part I. The structure of the yellow compound. Agr. Biol. Chem. 30, 232 (1966 a).
—— —— —— The photochemical reaction of pentachlorophenol. Part II. The chemical structures of minor products. Agr. Biol. Chem. 30, 239 (1966 b).

MITCHELL, L. C.: The effect of ultraviolet light (2537 A) on 141 pesticide chemicals by paper chromatography. J. Assoc. Official Agr. Chemists 44, 643 (1961).

NECKERS, D. C.: Mechanistic organic photochemistry. New York: Rheinhold (1967).

NIJHOFF, D. E., and E. HAVINGA: Photoreactions of aromatic compounds. VIII. A photochemical substitution of halogen in 2-bromo-(chloro)-4-nitroanisole. Tetrahedron Letters 47, 4199 (1965).

PENFOUND, W. T., and V. MINYARD: Relation of light intensity to effect of 2,4-dichlorophenoxyacetic acid on water hyacinth and kidney bean plants. Botan. Gaz. 109, 231 (1947).

PLIMMER, J. R.: Photolysis of amiben. Weed Sci. Soc. Amer. Abstr. p. 76, Washington, D.C. (1967).

——, and B. E. HUMMER: Photodecomposition of 2,3,6-trichlorobenzoic acid. Weed Sci. Soc. Amer. Abst. p. 20, New Orleans, La. (1968 a).

—— —— Photochemistry of herbicides: A study of some chlorinated compounds. 155th Amer. Chem. Soc. Nat. Meeting, San Francisco (1968 b).

—— —— Unpublished data (1968 c).

—— —— Photolysis of amiben (3-amino-2,5-dichlorobenzoic acid) and its methyl ester. J. Agr. Food Chem. 17, 83 (1969).

——, and U. I. KLINGEBIEL: Unpublished data (1968).

ROSEN, J. D., and R. F. STRUSZ: Photolysis of 3-(p-bromophenyl)-1-methoxy-1-methylurea. J. Agr. Food Chem. 16, 568 (1968).

—— ——, and J. D. STILL: Further studies concerning the photolysis of urea herbicides. J. Agr. Food Chem. (in press) (1969).

SANTHANAM, M., and V. RAMAKRSHNAN: Photosensitized oxidation of aniline. Indian J. Chem., pp. 88–90 (1968).

SHEETS, T. J.: Photochemical alteration and inactivation of amiben. Weeds 11, 186 (1963).

SLADE, P.: Photochemical degradation of paraquat. Nature 207, 515 (1965).

SMITH, G. N.: Ultraviolet light decomposition studies with Dursban[R] and 3,5,6-trichloro-2-pyridinol. J. Econ. Entomol. 61, 794 (1968).

SZYCHLINSKI, J.: Wstepne badania organicznych produktow fotolizy bromobenzenu w roztworze metanolowym. Zeszyty Nauk. Mat., Fiz., Chem., Wyzsza Skola Pedagog. Gdansk 3, 127 (1963).

——, and L. LITWIN: Badania nad fotochemia chlorowcopochodnych aromatycznych VI. Roczniki Chem. 37, 671 (1963).

TANG, C-S., and D. G. CROSBY: Photodecomposition of 3-(p-chlorophenyl)-1,1-dimethylurea (Monuron). 156th Amer. Chem. Soc. Nat. Meeting, Atlantic City, N.J. (1968).

TURRO, N. J.: Photochemical reactivity. J. Chem. Ed. 44, 536 (1967).

—— Molecular photochemistry. New York: Benjamin (1965).

UGOCHUKWU, E. N., and R. L. WAIN: Photolysis of 3,5-diiodo-4-hydroxy-benzonitrile (Ioxynil) as a factor in its herbicidal action. Chem. Ind. (London), p. 35 (1965).

WALCZAK, M., and J. SZYCHLINSKI: Kinetya fotolizy isomerow kwasu chlorobenzoesowego. Zeszyty Nauk. Mat., Fiz., Chem., Wyzsza Skola Pedagog. Gdansk. 3, 117 (1963).

Weed Society of America. Herbicide handbook. Geneva, N.Y.: Humphrey Press (1967).

WELDON, L. W., and F. L. TIMMONS: Photochemical degradation of diuron and monuron. Weeds 9, 111 (1961).

WOLF, W., and N. KHARASCH: Photolysis of iodoaromatic compounds in benzene. J. Org. Chem. 30, 2493 (1965).

WIGHT, W. L., and G. F. WARREN: Photochemical decomposition of trifluralin. Weeds 13, 329 (1965).

ZIFFER, H., and N. SHARPLESS: Extension of the Hammett equation to photochemical quantum yields. J. Org. Chem. 27, 1944 (1962).

Tolerances of pesticide residues in Czechoslovakia

By

V. Beneš * and V. Černá *

Contents

I. Introduction

The increasing application of chemicals in agriculture, in which chemical protection of plants forms an important part, called for a number of precautions to protect the workers in production and application as well as to decrease and possibly exclude residue risks with the consumers. One of the significant precautions, similar as in other countries, is the establishment of tolerances for pesticide residues. Since 1965, tolerances form part of the positive list of pesticide residues. Until 1968 tolerances had to be approved by the Chief Hygienist of the ČSSR; since 1969, when the State was divided into two Federal States, approval falls under the jurisdiction of the Chief Hygienists of both Republics.

---

* Institute of Hygiene, Prague.

The special data involved are evaluated by the Committee for Pesticides, this being the advisory council of the Chief Hygienist. This task is being carried out gradually on the basis of our own as well as of foreign results of research. The approved tolerances are published in the form of decrees by the Chief Hygienist (1965, 1966, and 1968) for organs of the Hygiene Service in regions and districts under whose competence tolerances supervision falls. When recommending tolerances, the Committee for Pesticides proceeds on conception mainly based on the acceptable daily intake (ADI) according to the Joint Committee of Experts (WHO/FAO 1962) for pesticide residues. The tolerances represent if possible the lowest achievable and necessary limit in agriculture with a certain margin covering conditions of applications so that they are simultaneously criteria for adhering to them (Beneš 1962, Beneš and Černá 1967).

II. The positive list of residues

The positive list of pesticide residues has been for practical reasons divided into four parts. Tolerances are stated in the first and second part for pesticides applied during vegetation (excepting herbicides) as well as for pesticides applied after harvest, respectively. The third part contains herbicides the application of which is stipulated by precise pre-emergence conditions in the early stage of vegetation. For the time being we do not consider the necessity of establishing their tolerances. Proposal of tolerances is, however, being considered for relatively persistent herbicides in some foodstuffs. The fourth part contains insecticides and fungicides applied to limited extent during vegetation in which also establishment of tolerances is considered unnecessary for the following reasons: a) residues do not occur at the time of harvest at a higher level than the required sensitivity of determination amounts to, whereby these values may be on the basis of present toxicological information considered to be negligible; b) residues are proved to form an unimportant part of the ADI; and c) pesticides are not applied to the edible part of products, e.g., trees in the period of vegetative rest or before blooming, seed dressing, etc. This part of the list also contains dressings on the basis of organically bound mercury. Mercury residues in consumed foods must not exceed the normal background which has been in foods for a number of years. Heptachlor is included in the positive list for seed dressing of sugar beet in combination with thiram.

This part of the list includes also fungicides containing coloidal or ground sulphur and calcium polysulphides, application of which is limited to protection periods lasting three to seven days.

III. The Methodical Manual of Plant Protection

A further important measure in practice, which regulates the application of pesticides, is afforded by the Methodical Manual of Plant

Table I. *Positive list of pesticide residues. Part I* [a]

General term	Effective substance	Tolerance (mg./kg.)	Product
Cupric compounds	Copper oxide and cupric oxide chloride	10 [b]	Fruit, vegetables, beet leaves
DDT [c]	2,2-bis(4-chlorphenyl)-1,1,1-trichloroethane	1	Apples, pears, plums, morello, cherries, vegetable leaves
Dichlorvos	O,O-Dimethyl-O-(2,2-dichlorvinyl) phosphate	0.1	Fruit, vegetables
Fenitrothion	O,O-Dimethyl-O-(3-methyl-4-nitrophenyl) thiophosphate	0.5	Fruit, vegetables
Lindane	1,2,3,4,5,6-hexachlorcyclohexane, gamma isomer	2.0	Fruit, vegetable leaves
Malathion	O,O-Dimethyl-S-(1,2-discarbethoxyethyl) dithiophosphate	1.0	Fruit, vegetables
Mevinphos	2-Carbomethoxyl-1-methyl-vinyldimethyl phosphate	0.1	Vegetable leaves, cabbage stalks
Thiomethon	O,O-Dimethyl-S-(2-ethylthioethyl) dithiophosphate	0.5	Fruit, winter cabbage, clover, lucerne, beet leaves
Trichlorphon	O,O,-Dimethyl-2,2,2-trichloro-1-hydroxyethyl phosphonate	1.0	Fruit, vegetable leaves
Zineb	Zinc ethylenebisdithiocarbamate	2.0	Fruit, vegetables

[a] Under preparation: Phormothion and amidithion (cherries), captan and pholpet (fruit, wine grapes).
[b] As Cu.
[c] Dusts, wettable powders, aerosols permitted.

Table II. *Positive list of pesticide residues. Part II* [a]

Name	Trade name	Tolerance (mg./kg.)	Product
Hydrogen cyanide		10.0	Cereal, leguminous plants
		3.0	Flour
		1.0	Kernel fruit
Hydrogen phosphide	Phostoxin	0.01	Corn (grain)
	Delicia	0.01	Rice, unroasted coffee, peanuts for roasting
Methyl bromide		50.0 [b]	Cereals, leguminous plants, cocoa beans, kernel fruits, unroasted coffee
		100.0	Spices
		200.0	Dried mushrooms
Pyrethrins	} Pybuthrine {	—	Corn (grain) [e]
Piperonyl butoxide		—	Packed food

[a] Under preparation: malathion (corn), diphenyl (lemons)
[b] As total bromide.
[e] After storing three months.

Protection, issued yearly by the Ministry of Agriculture and Nutrition and which has to be approved by the CHIEF HYGIENIST (1962–1967, 1968). It also contains a complete list of pesticides approved for the respective year with stipulated conditions of application. By means of these methodical instructions limited application is realized in cases when, for example, there are no complete hygienico-toxicological data for the estimation of the ADI at disposal, but if, on the other hand, there are sufficient data to exclude the risks of residues at the given limited application. When considering the problems of residues it is important that the Methodical Manual contains, for individual plants, their vegetation period and their pests with always certain permitted pesticides. The periods of protection and the listing of pesticides are liable to approval by the Chief Hygienist as they are closely connected with the risks of residues. Other pesticides with other effective substances, other formulations, or other concentrations can not be applied and this makes possible easier control of residues.

In order to decrease the risks of residues there was also introduced a survey on the individual fields treated by pesticides which is recorded directly by the agricultural enterprise. Besides the usual brief data it also contains records on neighbouring cultures. Correct survey makes most controls of residues unnecessary, but it affords rapid solution if neighbouring cultures are affected, it simplifies the work of organs of the Hygiene or Veterinary Service, and it makes it possible to react to some unforeseen circumstances and thus facilitates connection between research and practice.

IV. Current programs

We shall mention some findings by our reference laboratory, laboratories of the Regional Hygiene Service, and other institutes which, along with the results of research works, establish one of the bases for solving the hygienico-toxicological aspects of residues. At present we are dealing with the problem of unintentional residues, for which is being considered the use of the expression "practical residue limits" similarly as used in "Codex Alimentarius". This approach would comply best with the conception of our positive list. For these purposes there are being compiled results of determining actual intake of residues, especially DDT in foods of animal origin. Simultaneously the regulations for decontamination of food, forage, agricultural buildings, storerooms, etc. are being revised in order to limit superfluous sources of unintentional residues.

a) DDT

DDT residues in the periods of protection lasting 30 days after practical application of wettable powder on fruit and vegetables at the time of harvest are below one p.p.m. After application of DDT aerosol on cherries after three weeks 0.1 p.p.m. were determined (REFERENCE LABORATORY 1958–1968). Emulsion formulations are not permitted as the results yield higher residue levels. DDT is being substituted by organophosphorus insecticides so that only a small part of the daily intake is at present formed by intentional DDT residues, especially if the normal hygienic principles of washing fruit and vegetables before consumption are being adhered to. "Unintentional" DDT residues form the basic part of actual daily intake.[1]

In order to find out the daily intake of DDT residues, for a period of three years in four districts there were analyzed luncheons in communal catering (SALČÁK et al. 1969). Due to restricting measures, the mean content of residues appears to have a decreasing tendency from 1964 to 1966 in Czech regions, amounting to 0.22, 0.1, and 0.07 mg./luncheon portion, respectively. The DDT residue intake from one luncheon corresponded to 0.003, 0.0014, and 0.001 mg./kg. of weight, respectively, over this three year period. The luncheon represents approximately one third of the total daily food. Under this assumption it is possible to estimate in the studied representative group an approximate three-year daily intake of DDT of 0.009, 0.0042, 0.003 mg./kg. of weight, respectively.

When DDT residues in milk and butter were followed up for a period of one year once a month in six to eight dairies this disclosed in

---

[1] DDT is not used for fruit treatment since 1970.

15.6 percent of the samples values in the whole milk of 0.008 to 0.013 p.p.m; in butter were in 25 percent of the studied samples levels in the range of 0.124 to 0.220 p.p.m. and in beef tallow in 84 percent of samples 0.048 to 0.336 p.p.m. (HRUŠKA 1969). For control determinations, the well-known Schechter-Haller spectrophotometric method was applied. The determination of DDT and DDE residues in butter by means of gas chromatography shows in Slovakia the levels of 0.07 to 0.9 p.p.m for DDT and 0.1 to 0.62 p.p.m. for DDE (UHNAK *et al.* 1969).

b) Fenitrothion

In the Czechoslovak product Metation E-50, the residue decrease in fruit, vegetables, and beetroot was studied by a spectrophotometric method. The residues decrease to a value of 0.5 p.p.m at intervals of two to 12 days in tomatoes, capsicum, cabbage, sprouts, strawberries, cherries, pears, turnips, and cauliflower. Greater persistence appeared in apples and plums, from 14 to 21 days (BÁTORA and KOVÁČ 1965, ČERNÁ 1968). In Brussels sprouts after 35 days were found 2.6 p.p.m. on the leaves whereas the edible part contained 0.05 p.p.m. However this field experiment was carried out in colder autumn months (ČERNÁ and BENEŠ 1968 a).

c) Malathion

When controlling tomatoes and capsicum in the field, the residues determined by a spectrophotometric method were smaller than 0.2 p.p.m. ten days after application (REFERENCE LABORATORY 1958–1968). In lettuce the residue after 15 days did not exceed one p.p.m. as determined by colorimetric and enzymatic methods (ČERNÁ and BENEŠ 1961).

d) Dichlorvos

After application on plums, apples, cabbage, and cauliflower, residues after three to six days were proved in the amount of 0.1 p.p.m. (ČERNÁ and BENEŠ 1969).

e) Mevinphos

After three days residues on cauliflower were determined at 0.1 p.p.m. even after application in cool autumn months (MARTINCOVÁ and CÍTKOVÁ 1969).

f) Trichlorphon

In cherries after six days residues reached the level of 0.8 p.p.m. colorimetrically (BENEŠ *et al.* 1963).

g) Zineb

Ten days after application on tomatoes and capsicum residues did not exceed two p.p.m. (REFERENCE LABORATORY 1958–1968).

h) Hydrogen cyanide

Numerous analyses of grain samples (wheat, barley, rye) disclosed after fumigation with hydrogen cyanide in doses up to one kg./100 cu. m., after leaving stores, residues within the limits of two to six p.p.m After fumigation of flour in doses of one kg./100 cu. m. after two to three days daily airing, the values in wheat flour are mostly within limits of one to 5 p.p.m.; in rye flour findings are higher, up to 10 p.p.m. Further airing causes residues to decrease and analyses carried out eight to nine days later proved values below one p.p.m. In order to find out the actual hydrogen cyanide intake in treated flour, bread was experimentally baked. Residue contents of the flours were within the limits three to nine p.p.m.; in the baked bread residues were from 0.1 to 0.4 p.p.m. and in only one case did they reach one p.p.m. Fumigation of fresh kernel fruit (apples, pears) is only carried out exceptionally in imports at the border as protection against San Chosé worm; in market control the residue values in apples amounted to less than one p.p.m.

At present the possibilities of hydrogen cyanide application on some oil seed plants are being studied. In laboratory experiments residues after hydrogen cyanide application in rape seed, poppy seed, sunflower seed, and shelled peanuts were followed up. After three days of airing the following values were determined: 44.5, 108.4, 35.2, and 18.2 p.p.m., respectively; after 20 days of airing the values were: 11.3, 21.3, 24.2, and 2.8 p.p.m., respectively (REFERENCE LABORATORY 1958–1968, LISTOPADOVÁ et al. 1969).

i) Hydrogen phosphide

Values found one to two months after treatment of grain with Phosphotoxin were within the limits 0.02 to 0.05 p.p.m. In single cases values were up to 0.08 p.p.m. After cleaning residues do not exceed 0.01 p.p.m. After fumigation by preparation Delicia these values 0.01 p.p.m. are not exceeded even without cleaning the grain.

There is a tendency to use phosphine to a wider extent in the form of preparation Delicia and for this reason the technological procedure for raw oil seed and coffee beans was experimentally tested. In green coffee in three days of exposure after ten days of airing, phosphine residues were within the limits 0.021 to 0.036 p.p.m. After roasting this coffee, no residues were found at the limiting detectability of the method at 0.004 p.p.m.

Phosphine residues in sunflower seed, rape seed, poppy seed, and shelled peanuts after two days of airing were 0.012, <0.004, <0.004, and 0.056 p.p.m., respectively; after 19 days of airing they were 0.005, <0.004, <0.004, and 0.033 p.p.m., respectively (Černá 1960, Listopadová et al. 1969).

j) Methyl bromide

According to numerous findings by control laboratories bromide ion residues including natural content amount in wheat to 20 to 41 p.p.m. and in rice 11 to 28 p.p.m. In rice samples with proved double treatment there were, however, much higher residues proved (86 p.p.m.). Among leguminous plants, fumigation is mostly carried out in peas in which in the period of marketing values of eight to 28 p.p.m. were found, in crushed peas (powder) 20 p.p.m. In peeled almonds values from ten to 42 p.p.m. were determined. In shelled unroasted peanuts findings usually amount to approximately 100 p.p.m. and more. Long-period airing or roasting did not decrease these values to any extent. In spices (pepper, marjoram, red pepper, cinnamon, caraway, and bay-leaf) residues amounted to 20 to 90 p.p.m. Values above 100 p.p.m. were determined only in single cases. In dried mushrooms after three days of airing values of 75 to 200 p.p.m. were determined (Reperence Laboratory 1958–1968, Listopadová et al. 1969).

V. Conclusions

In consideration of the rather difficult toxicological evaluation and analysis of methyl bromide residues as such, although in practice often wider application of methyl bromide for fumigation of plants is required, we are in favour of studying further procedures by means of hydrogen cyanide and hydrogen phosphide as this opens up perspective.

In hydrogen cyanide, as the mentioned findings disclose, the actual daily intake of residues after fumigation of cereals and flour and its culinary treatment is minimal, so that in consideration of the maximum acceptable daily intake of 0.01 mg./kg. body weight a sufficient reserve is afforded to exclude the risks of residues.

In hydrogen phosphide unwanted chemical absorption is improbable. Residues in food treated by Delicia do not exceed 0.01 p.p.m., disregarding further culinary treatment of foods by cooking, etc. The sensitivity of the method may reach up to 0.002 p.p.m. (Černá 1960) and in this case it may be—from the toxicological point of view—considered the criterian for negligible residues. As far as analytical methods used in practice for residue control are concerned adequate methods were recommended and approved for a number of pesticide residues in plant material taking into consideration the apparatus and other equipment of control laboratories. These methods are meant for the determination of residues of DDT, lindane, organophosphates, and fumigants. In practice, however, any analytical methods of sufficient accuracy and reproducibility may be used.

Summary

Tolerances of pesticide residues, approved by the Chief Hygienist of the ČSSR as proposed by the Commission for Pesticides in the years 1965 to 1968, are listed in the first and second part of the positive list of allowable pesticide residues, which contains pesticides approved for application during vegetation and for protection of foodstuffs after harvest. Herbicides, the other insecticides, and fungicides in which in view of conditions of application it is not considered necessary to determine tolerances, are listed in the third and fourth parts of the list. Conditions of application are published each year in the Methodical Handbook on Plant Protection, which is liable for approval by the Chief Hygienist. Hygienico-toxicological evaluation of residues is based on the conception of the Joint Committee of Experts FAO/WHO. The tolerance is, if possible, determined as the lowest limit taking the essential demands for agricultural practice into consideration.

Stated are findings serving for solution of questions of unintentional DDT residues and some results of the control of application procedures in field practice with fenitrothion, malathion, mevinphos, dichlorvos, trichlorphon, zineb, hydrogen cyanide, hydrogen phosphide, and methyl bromide are demonstrated.

Résumé *

Tolérances concernant les résidus de pesticides en Tchécoslovaquie

Les tolérances concernant les résidus de pesticides qui ont été approuvées par l'Hygiéniste en Chef de la ČSSR sur proposition de la Commission des Pesticides au cours des années 1965 à 1968 figurent dans la première et la deuxième partie de la liste positive des résidus de pesticides permissibles, qui comprend les pesticides dont l'application est approuvée durant la période de croissance des végétaux et pour la protection des stocks après la récolte. Les herbicides, les autres insecticides et les fongicides pour lesquels, en raison des conditions d'application, il n'apparaît pas nécessaire de fixer des tolérances sont énu-

---

* Traduit par S. DORMAL-VAN DEN BRUEL.

mérés dans la troisième et la quatrième partie de la liste. Les conditions d'application sont publiées annuellement dans le Manuel des Méthodes pour la Protection des Plantes qui est soumis pour approbation à l'Hygiéniste en Chef. L'évaluation hygiénique et toxicologique des résidus de pesticides est basée sur les principes du Comité Mixte d'Experts FAO/OMS. La tolérance est établie, si possible, à la limite la plus basse qui soit compatible avec les exigences essentielles de la pratique agricole.

Il est fait état d'observations destinées à résoudre le problème des résidus non intentionnels de DDT et de certains résultats de contrôle des procédés d'application pratique du fénitrothion, du malathion, du mevinphos, du dichlorvos, du trichlorphon, du zinèbe, de l'acide cyanhydrique, de l'hydrogène phosphoré et du bromure de méthyle.

Zusammenfassung *

Toleranzen von Pestizidrückständen in der Tschechoslowakei

In den Teilen I und II der positiven Liste erlaubter Pestizidrückstände sind Toleranzen für Pestizidrückstände aufgeführt, die vom Haupthygieniker der ČSSR genehmigt wurden. Die Genehmigungen durch den Haupthygieniker erfolgen auf der Basis der 1965–68 von der Komission für Pestizide ausgearbeiteten Vorschläge. Die Liste enthält Pestizide, die zur Anwendung während der Vegetationszeit und zum Schutz der Ernte während der Lagerung zugelassen sind. In den Teilen III und IV der Liste sind die Herbizide und die anderen Insektizide und Fungizide aufgeführt, für welche in Anbetracht der Bedingungen ihrer Anwendung Toleranzen nicht festgelegt zu werden brauchen. Die Anwendungsbedingungen werden alljährlich im Methodical Handbook on Plant Protection (Methodisches Handbuch über Pflanzenschutz) veröffentlicht. Die Veröffentlichung muß vom Haupthygieniker genehmigt werden. Die hygienisch-toxikologische Bewertung der Rückstände basiert auf den Vorstellungen des Joint Committee of Experts FAO/WHO (des Experten-Ausschusses der Welternährungsorganisation/Weltgesundheitsorganisation). Als Toleranz wird, unter Berücksichtigung der Erfordernisse der landwirtschaftlichen Praxis, der unterste mögliche Grenzwert festgelegt.

Die Ergebnisse von Untersuchungen werden geschildert, die zur Lösung von Fragen dienen sollen, wie sie sich aus der Anwesenheit unbeabsichtigter DDT-Rückstände ergeben. Ferner werden einige Resultate der Überprüfung von Freiland-Anwendungsverfahren für Fenitrothion, Malathion, Mevinphos, Dichlorvos, Trichlorphon, Zineb, Blausäure, Phosphorwasserstoff und Methylbromid vorgeführt.

* Übersetzt von H. MAIER-BODE.

References

Bátora, V., and J. Kováč: Personal communication (1965).

Beneš, V.: Entwurf eines positiven Verzeichnisses von Pflanzenschutz mitteln in der Tschechoslowakischen Sozialistischen Republik. Tagungsber. DAL Berlin 42, 69 (1962).

——, and V. Černá: Pflanzenschutz vom Gesichtspunkt der Pestizidrückstände in der Tschechoslowakischen Sozialistischen Republik. Proc. 7th Internat. Congress Nutrition, Hamburg (1966); Problems of World Nutrition, vol. 4, p. 700. New York: Pergamon Press (1967).

——, L. Rosíval, and V. Černá: Prevention of pesticides in food. Cs. Hyg. 8, 98 (1963).

Černá, V.: Residues of hydrogen phosphide in grain treated with "Phostoxin." Cs. Hyg. 5, 311 (1960).

—— A study on organophosphate insecticides—Possible food contaminants. Ph.D. Thesis, Inst. Chem. Technol., Prague (1968).

——, and V. Beneš: Hygienic evaluation of Fosfothion residues. Cs. Hyg. 6, 149 (1961).

—— —— Unpublished data (1968).

—— —— Residues of dichlorvos after application on fruit and vegetables. Cs. Hyg. 14, 113 (1969).

Hruška, J.: DDT residues in milk, butter, and beef tallow. In: Pesticide residues in crops. Czechoslovak Agr. Acad., Inst. Scientific Tech. Information, Prague, No. 71 (1969).

Listopadová, E., A. Klein, and S. Gielerová: Fumigant residues in foods. In: Pesticide residues in crops. Czechoslovak Agr. Acad., Inst. Scientific Tech. Information, Prague, No. 71 (1969).

Martincová, J., and M. Čitková: Following up of Metation and Phosdrin in plant material. In: Pesticide residues in crops. Czechoslovak Agr. Acad., Inst. Scientific Tech. Information, Prague, No. 71 (1969).

Methodical Manual on Plant Protection, Ministry of Agr. and Nutrition, Prague (1968).

Methods of Plant Protection, Ministry of Agr., Forestry, and Water Supplies, Prague (1962–1967).

Positive List of Pesticide Residues. Decree of Chief Hygienist of the ČSSR: HE-343.3-9.7.65 of July 12, 1965; HE-343.3-1.7.66 of July 15, 1966; HE-343.3-28.12.68 of Dec. 28, 1968.

Reference Laboratory, Inst. of Hygiene, Prague: Unpublished data (1958–1968).

Salčák, J., J. Koblížek, and J. Sponař: Studying of DDT residues in foods of selected establishments of communal feeding. In: Pesticide residues in crops. Czechoslovak Agr. Acad., Inst. Scientific Tech. Information, Prague, No. 71 (1969).

Uhnak, J., A. Szokolay, and F. Görner: Einige Ergebnisse der Untersuchungen von chloroganischen Insektizidrückständen in den Lebensmitteln. Proc. 8th Internat. Congress Nutrition, Prague, Abstr. U-1 (1969).

WHO/FAO: Principles governing consumer safety in relation to pesticide residues. World Health Org., Tech. Rep't. Series 240 (1962).

Terminal residues of pyrethrin-type insecticides and their synergists in foodstuffs

By

Joseph B. Moore *

Contents

I. Introduction

Pyrethrins and synergists are usually used to protect food products in postharvest application (LINDGREN *et al.* 1968). Data collected for food additive petition give us reliable information on the level of residues from synergized pyrethrin sprays. The data were obtained on the synergists by the gas chromatographic method of BRUCE (1967). The data on the pyrethrins residue were by a gas chromatographic method of BRUCE as yet unpublished.

II. Residues in fatty foods

Tests to determine residues in "fatty" foods from pyrethrins and synergist sprays in warehouse or food processing situations were set up in a model room as follows:

---

* McLaughlin Gormley King Co., Minneapolis, Minn.

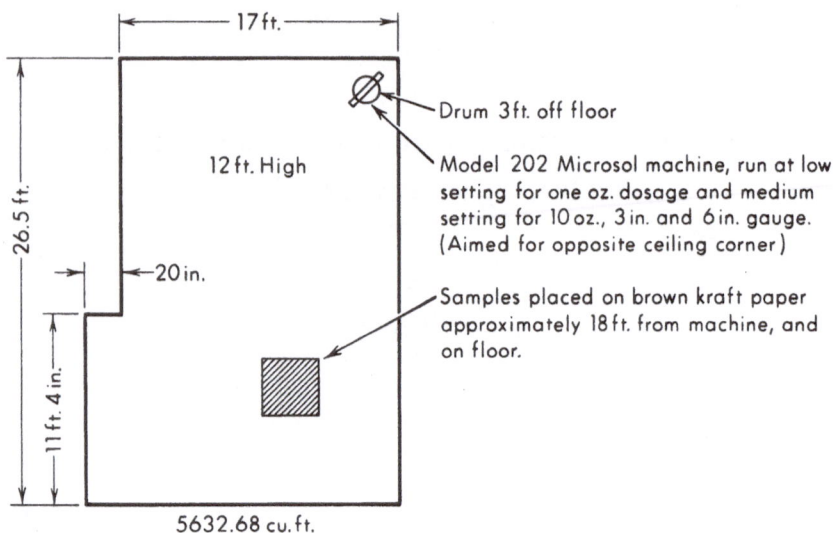

Drum 3ft. off floor

Model 202 Microsol machine, run at low setting for one oz. dosage and medium setting for 10 oz., 3 in. and 6 in. gauge. (Aimed for opposite ceiling corner)

Samples placed on brown kraft paper approximately 18 ft. from machine, and on floor.

Spray composition:	0.50% Pyrethrins 1.00% Piperonyl butoxide 1.67% MGK 264 96.83% Petroleum distillate (Dispersol)
Dosages:	One fluid oz./1,000 cu. ft. or 5.6 fluid oz. for this room Ten fluid oz./1,000 cu. ft. or 56 fluid oz. for this room
Commodities:	Duplicate samples for each test and each dosage: *Bacon, sliced*—covered with one thickness of polyethylene and uncovered to get full drop-out; approximately 0.25 lb./sample *Margarine*—covered in original package—one thickness of wax paper and outer box of heavy waxed paper; also, duplicate samples completed uncovered to catch full drop-out; 0.25 lb. stick/sample *Brazil nuts, shelled*—covered in heavy pliofilm in unbroken original package; duplicate samples open to complete fall-out; samples approximately three oz. each

Miscellaneous conditions: Temperature approximately 80° F. during test. At one oz. dosage, spray allowed to settle for ½ hour. Samples sealed in pliofilm bags and put in deep-freeze within two hours of conclusion of spraying. At ten oz. dosage, four hours allowed for spray settling. At this time air was completely clear and margarine and bacon were running some. Samples sealed in pliofilm bags and placed in deep-freeze approximately five hours after end of spraying.

Results of this test are shown in Table I.

The data shown in Table I indicate the type of residues expected from treatment of warehouse and storage areas or indirect contamination as opposed to postharvest treatment directly on the food product.

III. Residues in dried codfish

Using the same analytical procedures, new data have been obtained from direct or postharvest applications of pyrethrins and synergists. Synergized pyrethrins are used to protect cod fillets from flies during sun drying. Usually an emulsion containing 0.0625 percent pyrethrins and 0.1250 percent piperonyl butoxide is used as a dip. The dip time is four to five seconds.

Samples of dried cod from The Newfoundland Associated Fish Exporters, Ltd., were sampled about 39 days after dipping.

The residues from this treatment showed levels of 0.056 to 0.0661 p.p.m. of pyrethrins and 0.8675 to 1.100 p.p.m. of piperonyl butoxide in four replicates.

IV. Residues in potato chips

Synergized pyrethrins are used in the potato chip industry. The potatoes are stored in warehouses in wooden crates. Spray or fog of synergized pyrethrins is used to protect the potatoes in storage from the ravages of the potato tuber moth and sciara flies.

A series of tests was set up in cooperation with the Wise Potato Chip Corporation to determine what, if any, residue problem there might be on potato chips from such treatment.

The treatments used were as shown in Table II.

Table I. *Residues in fatty foods*

Product analyzed	Treatment dosage (oz.)	Residue (p.p.m.)			
		Pyrethrins		MGK 264	Piperonyl butoxide
		I	II		
Oleomargarine (wrapped)	10	0.025	0.071	0.516	0
	10	0.026	0.059	0.397	0
Oleomargarine (unwrapped)	10	1.565	1.048	7.738	4.070
	10	1.663	1.048	7.500	3.450
Bacon (wrapped)	10	0.020	0	0.437	0
	10	0.040	0.029	0.496	0
Bacon (unwrapped)	10	1.631	1.195	8.650	3.930
	10	1.739	1.363	9.680	3.790
Brazil nuts (packaged)	10	0	0	0	0
	10	0	0	0	0
Brazil nuts (exposed	10	2.283	2.306	15.950	7.590
	10	2.345	2.201	16.350	7.380
Oleomargarine	None	0	0	0	0
		0	0	0	0
Bacon	None	0	0	0	0
		0	0	0	0
Brazil nuts	None	0	0	0	0
		0	0	0	0
Oleomargarine (wrapped)	1	0	0	0	0
	1	0	0	0	0
Oleomargarine (unwrapped)	1	0.251	0.178	1.429	0.793
	1	0.257	0.178	1.746	0.759
Bacon (wrapped)	1	0	0	0	0
	1	0	0	0	0
Bacon (unwrapped)	1	0.235	0.210	2.575	0.966
	1	0.233	0.210	2.380	0.862
Brazil nuts (packaged)	1	0	0	0	0
	1	0	0	0	0
Brazil nuts (exposed)	1	0.234	0.249	2.282	0.931
	1	0.250	0.272	2.240	1.103

Table II. *Treatments for potato chip study*

Sample code	Treatment
1	Untreated, unpeeled control
2A	Fogged with 0.3% pyrethrins and 1.5% piperonyl butoxide in storage—normal treatment
2B	Fogged as above—2 × normal treatment.
2C	Fogged as above—4 × normal treatment
3	Potatoes fogged directly with 0.3% pyrethrins and 1.5% piperonyl butoxide
4	Peeled, untreated potatoes
5	Potato chips from peeled, untreated potatoes
6	Potato chips from peeled potatoes after being dipped in 0.3% pyrethrins and 1.5% piperonyl butoxide
7A	Unpeeled potatoes dusted with 0.2% pyrethrin dust at 1 lb./ton rate
7B	Unpeeled potatoes dusted as above at 5 lb./ton rate
8A	Potato chips from unpeeled potatoes dusted at 1 lb./ton rate
8B	Potato chips from unpeeled potatoes dusted at 5 lb./ton rate
9	Potato chips from unpeelel potatoes fogged directly with 0.3% pyrethrins and 1.5% piperonyl butoxide (see 3)
10	Peeled potatoes dipped directly in 0.3% pyrethrins and 1.5% piperonyl butoxide

Table III. *Analysis of potatoes for pyrethrins residues*

Code [a]	Injection		Peak height (mm.)	Pyrethrins (p.p.m.)
	g./ml.	μl.		
1S	4	2.5	0	0
1K	4	2.5	0	0
2AS	4	2.5	1.1	0.0016
2AK	4	2.5	1.5	0.0022
2BS	4	2.5	1.5	0.0022
2BK	4	2.5	3.5	0.0050
2CS	4	2.5	2.6	0.0037
2CK	4	2.5	4.3	0.0062
3S	4	2.5	9.7	0.0139
3K	4	2.5	15.0	0.0216
4S	4	2.5	0	0
4K	4	2.5	0	0
7AS	2	2.5	47.4	0.1362
7AK	2	2.5	76.0	0.2184
7BS	0.4	1.0	66.6	2.3920
7BK	0.4	1.0	127.6	4.5830
10S	4	2.5	62.2	0.0694
10K	4	2.5	62.0	0.0692

[a] See Table II.

Using the same method of analysis described above, the residues from these different treatments are shown in Tables III and IV.

With injections of 2.5 μl. extract containing four g. of potato residue/ml., one ng. peak (69.6 mm.) is equivalent to 0.1 p.p.m. of pyrethrins.

Table IV. *Analysis of potatoes for piperonyl butoxide residues*

| Code [a] | Injection | | Peak height (mm.) | Piperonyl butoxide (p.p.m.) |
	g./100 ml.	μl.		
1S	4	2	0	0
1K	4	2	0	0
2AS	4	2	3.9	0.0135
2AK	4	2	5.0	0.0174
2BS	4	2	7.4	0.0257
2BK	4	2	8.1	0.0281
2CS	4	2	9.1	0.0316
2CK	4	2	13.9	0.0483
3S	4	2	17.3	0.0600
3K	4	2	23.9	0.0830
4S	4	2	0	0
4K	4	2	0	0
7AS	4	2	—	—
7AK	4	2	—	—
7BS	4	2	—	—
7BK	4	2	—	—
10S	2	2	41.0	0.2850
10K	2	2	42.5	0.2950

[a] See Table II.

V. Residues in treated grain

HEAD et al. (1968) showed that piperonyl butoxide stablizes pyrethrins exposed to sunlight and air. These studies were made by exposing a thin film of pyrethrins and piperonyl butoxide on glass plates to sunlight and air. They used electron capture gas chromatography to determine the level of pyrethrins. This method was described by HEAD (1966).

HEAD in unpublished work has investigated the stability of pyrethrins on wheat exposed to sunlight and air. Table V shows the residue of pyrethrins (cinerins and pyrethrins) remaining on the wheat.

The 'pyrethrins' on wheat show much greater stability than in the thin film exposure of the untreated controls. The 'Pyrethrin II' components tend to be more stable than the 'Pyrethrin I' components and the cinerins are more stable than the pyrethrins.

Table V. *Cinerins and pyrethrins residues on wheat*

Sample	Residues (p.p.m.)							Loss total pyrethrins (%)
	Cin.I	Py.I	Cin.II	Py.II	'Py.I'	'Py.II'	Tot.	
Formulation	0.71	1.88	1.08	1.42	2.59	2.50	5.09	—
Control								
(unexposed)	0.66	1.81	1.08	1.35	2.47	2.43	4.90	—
10-hr. exp.	—	—	0.006	0.002	—	0.008	0.008	99.8
15-hr. exp.	—	—	0.004	0.001	—	0.005	0.005	99.9
Wheat control	0.76	1.76	1.05	1.39	2.52	2.43	4.95	—
10-hr. exp.	0.24	0.41	0.51	0.57	0.65	1.08	1.73	65
15-hr. exp.	0.28	0.41	0.51	0.77	0.69	1.28	1.97	60
24-hr. exp.	0.19	0.20	0.23	0.28	0.39	0.51	0.90	81

Although the extract from the wheat is typical of degraded material, both pyrethrin I and pyrethrin II are still clearly present. It is probable, therefore, that when wheat is sprayed with 'pyrethrins', part of the active constituents are absorbed and protected whilst the material at the surface suffers rapid degradation.

The electron capture gas chromatographic tracings show that the degradation of pyrethrins on grain is typical of the degradation of pyrethrins' films on glass.

It is interesting to note that McLaughlin Gormley King Company had samples of the degraded material compared with the original extract for acute oral LD_{50} on rats. The LD_{50} of the original extract was 7.5 g./kg. of body weight versus eight g./kg. for the degraded material.

VI. Residues in chickens and eggs

Chickens are sometimes sprayed or fogged with synergized pyrethrins for control of ectoparasites. A rather severe test was set up to determine the extent of residues from pyrethrins and piperonyl butoxide in the chicken carcasses and eggs. These data are taken from an unpublished report from Niagara Chemical Division, FMC Corporation.

The chickens were dipped bodily twice weekly for seven weeks in emulsions of pyrethrins and piperonyl butoxide.

The treatments consisted of TI (an emulsion containing 0.085 percent pyrethrins and 0.85 percent piperonyl butoxide) and TII (an emulsion containing 0.85 percent pyrethrins and 8.5 percent piperonyl butoxide) and the untreated group.

The results of these analyses are shown in Tables VI through XIII.

Table VI. *Analysis of eggs for pyrethrin I resulting from
dipping chickens in pyrenone, two concentrations*

Treatment [a]	Weeks	Injection (μl.)	Peak height (mm.)	Pyrethrins in eggs (p.p.m.)
TI	1	2	1.2	0.0097
TI	2	2	7.5	0.0608
TI	3	2	11.2	0.0908
TI	4	2	17.0	0.1379
TI	5	2	23.4	0.1898
TI	6	2	20.6	0.1671
TI	7	2	27.8	0.2255
0	8	2	10.8	0.0876
0	9	2	6.0	0.0487
0	10	2	5.4	0.0438
0	11	2	3.3	0.0268
TII	1	2	24.7	0.2003
TII	2	2	41.3	0.3350
TII	3	2	52.1	0.4225
TII	4	2	54.0	0.4380
TII	5	2	65.1	0.5280
TII	6	2	72.3	0.5864
TII	7	2	80.4	0.6521
0	8	2	45.2	0.3666
0	9	2	29.1	0.2360
0	10	2	25.0	0.2028
0	11	2	20.8	0.1687

[a] See text.

Table VII. *Analysis of eggs from untreated chickens*

Week	Pyrethrins [a]		Piperonyl butoxide [b]	
	Injection (μl.)	Peak height (mm.)	Injection (μl.)	Peak height (mm.)
1	2	0	1	0
2	2	0	1	0
3	2	0	1	0
4	2	0	1	0
5	2	0	1	0
6	2	0	1	0
7	2	0	1	0
8	2	0	1	0
9	2	0	1	0
10	2	0	1	0
11	2	0	1	0

[a] Extractives from 50 g. of eggs in 50 ml. of hexane. With two μl. injections, a peak height of 61.65 mm. (one ng. of pyrethrins) is equivalent to 0.5 p.p.m.

[b] Extractives from 25 g. of eggs in 100 ml. of hexane. With one μl. injection, a peak height of 36.5 mm. (one ng.) is equivalent to four p.p.m. of piperonyl butoxide.

Table VIII. *Analysis of eggs for piperonyl butoxide resulting from dipping chickens in pyrenone, two concentrations*

Treatment [a]	Weeks	Injection (μl.)	Peak height (mm.)	Piperonyl butoxide in eggs (p.p.m.)
TI	1	1	3.3	0.362
TI	2	1	7.1	0.779
TI	3	1	8.4	0.922
TI	4	1	12.3	1.349
TI	5	1	14.1	1.547
TI	6	1	16.4	1.799
TI	7	1	16.1	1.767
0	8	1	14.4	1.580
0	9	1	9.7	1.064
0	10	1	6.0	0.658
0	11	1	5.0	0.549
TII	1	1	23.5	2.579
TII	2	1	31.6	3.468
TII	3	1	34.5	3.787
TII	4	1	36.5	4.005
TII	5	1	38.5	4.225
TII	6	1	37.3	4.093
TII	7	1	39.8	4.367
0	8	1	32.4	3.555
0	9	1	29.6	3.248
0	10	1	25.3	2.776
0	11	1	20.5	2.249

[a] See text.

Table IX. *Analysis of chicken tissues for pyrethrins, lower concentration*

Tissue	Treatment [a]	Weeks	G. tissue/ 50 ml.	Injection (μl.)	Peak height (mm.)	Pyrethrins (p.p.m.)
Body fat	TI	7	2.5	4	24.2	1.594
		11	2.5	4	2.4	0.158
Skin	TI	7	1.25	4	41.5	5.468
		11	1.25	4	4.5	0.712
Muscles	TI	7	25	4	20.8	0.137
		11	25	4	2.2	0.015
Liver	TI	7	25	4	12.3	0.081
		11	25	4	3.3	0.021
Gizzard	TI	7	25	4	20.4	0.134
		11	25	4	5.2	0.034

[a] See text.

Table X. *Analysis of chicken tissues for pyrethrins, higher concentration*

Tissue	Treatment [a]	Weeks	G. tissue/ 50 ml.	Injection (μl.)	Peak height (mm.)	Pyrethrins (p.p.m.)
Fat	TII	7	2.5	4	73.3	4.829
		11	2.5	4	41.1	2.708
Skin	TII	7	0.125	4	45.1	59.420
		11	0.5	4	22.4	7.378
Muscle	TII	7	25	4	61.0	0.402
		11	25	4	24.6	0.162
Liver	TII	7	25	4	19.8	0.130
		11	25	4	3.9	0.024
Gizzard	TII	7	25	4	92.8	0.611
		11	23	4	29.3	0.193

[a] See text.

Table XI. *Analysis of untreated chicken tissues*

Tissue	Weeks	Pyrethrins			Piperonyl butoxide		
		G. tissue/ 50 ml.	Injection (μl.)	Peak height (mm.)	G. tissue/ 100 ml.	Injection (μl.)	Peak height (mm.)
Fat	7	2.5	4	0	5	2	0
	11	2.5	4	0	5	2	0
Skin	7	25	4	0	25	2	0
	11	25	4	0	25	2	0
Muscle	7	25	4	0	25	2	0
	11	25	4	0	25	2	0
Liver	7	25	4	0	25	2	0
	11	25	4	0	25	2	0
Gizzard	7	25	4	0	25	2	0
	11	25	4	0	25	2	0

Table XII. *Analysis of chicken tissues for piperonyl butoxide, lower concentration*

Tissue	Treatment [a]	Weeks	G. tissue/ 100 ml.	Injection (μl.)	Peak height [b] (mm.)	Piperonyl butoxide (p.p.m.)
Fat	TI	7	5	2	59.6	16.175
		11	5	2	11.0	2.985
Skin	TI	7	1.25	2	61.4	66.652
		11	1.25	2	8.0	8.685
Muscle	TI	7	25	2	26.0	1.411
		11	25	2	3.3	0.179
Liver	TI	7	25	2	25.6	1.390
		11	25	2	3.2	0.174
Gizzard	TI	7	25	2	38.2	2.073
		11	25	2	4.5	0.244

[a] See text.
[b] An 18.424 mm. peak height is equivalent to one p.p.m. with two μl. injection of a solution containing 25 g. of tissue extract/100 ml. of hexane.

Table XIII. *Analysis of chicken tissues for piperonyl butoxide, higher concentration*

Tissue	Treatment [a]	Weeks	G. tissue/ 100 ml.	Injection (μl.)	Peak height (mm.)	Piperonyl butoxide (p.p.m.)
Fat	TII	7	2.5	2	78.6	42.663
		11	2.5	2	38.4	20.843
Skin	TII	7	.125	2	52.9	574.250
		11	.125	2	10.8	117.239
Muscle	TII	7	12.5	2	42.9	4.657
		11	12.5	2	20.2	2.193
Liver	TII	7	12.5	2	31.5	3.420
		11	12.5	2	10.0	1.151
Gizzard	TII	7	12.5	2	52.2	5.606
		11	12.5	2	24.0	2.602

[a] See text.

The residues from these treatments indicate that no unusually high residues would occur in either the eggs or the chicken tissues by use of these sprays under the normal directions for use.

The concentration of material in TI would be about the amount recommended in a spray, but the dosage obtained from dipping is extremely high. The number of applications is also higher than would be experienced in actual use.

The severity of the test was intentional so as to reach residue levels that could be measured.

The very high levels in the skin include that material still on the outside of the skin.

VII. Conclusions

The recent development of suitable gas chromatographic methods for pyrethrins and piperonyl butoxide has enabled analysis to determine minute levels of these materials as residues on and in food products.

Further work is in progress and is needed in the area of identification and determination of levels of the breakdown products of these two chemicals.

Summary

This article on the terminal residues of pyrethrin-type insecticides and their synergists in foodstuffs gives recent data on residues in and on food products treated with these pesticides. These residues update previously published data. The residues are determined by more sensitive and accurate methods than previously.

Résumé *

Résidus des insecticides du type pyréthrine
et de leurs agents de potentialisation dans les matières alimentaires

Cet article sur les résidus des insecticides du type pyréthrine et de leurs agents de potentialisation donne des résultats récents intéressant les teneurs trouvées sur et dans les produits alimentaires traités par ces pesticides. Ces teneurs en résidus, déterminées par des méthodes plus sensibles et plus précises qu'auparavant permettent de réviser les résultats précédemment publiés.

Zusammenfassung **

Endrückstaende von Insektiziden des Pyrethrine-Typs
und ihre synergistischen Stoffe in Nahrungsmittel

Dieser Artikel über die Endrückstände von Insektiziden des Pyrethrine-Typs und ihre synergistischen Stoffe in Nahrungsmittel gibt neueste Daten über Rückstände in und auf Nahrungsmittel, die mit

---

* Traduit par R. Mestres.
** Übersetzt von M. Düsch.

diesen Pestiziden behandelt worden sind. Diese Rückstände bringen vorher veröffentliche Daten auf den neuesten Stand. Die Rückstände werden durch empfindlichere und genauere methoden als vorher bestimmt.

References

Bruce, W. N.: Detector cell for measuring picogram quantities of organophosphorus insecticides, pyrethrin synergists and other compounds by gas chromatography. J. Agr. Food Chem. 15, 178 (1967).

Head, S. W.: The quantitative determination of pyrethrins by gas liquid chromatography. Part 1: Detection by electron capture. Pyrethrum Post 8 (4), 3 (1966).

————, N. K. Sylvester, and S. K. Challinor: The effect of piperonyl butoxide on stability of crude and refined pyrethrum extracts. Pyrethrum Post 9 (3), 14 (1968).

Lindgren, D. L., W. B. Sinclair, and L. E. Vincent: Residues in raw and processed foods resulting from post-harvest insecticidal treatments. Residue Reviews 21, 1 (1968).

Acaricide residues on citrus foliage and fruits
and their biological significance

By

L. R. Jeppson * and F. A. Gunther *

Contents

I. Introduction

The unprecedented increase in the number and variety of pesticides developed during the past two decades has resulted in the availability of at least a dozen acaricides effective in the control of mites injurious to citrus. Although most of these acaricides effectively controlled the citrus red mite, *Panonychus citri* (McG.), the resistance phenomenon

* Department of Entomology, *University of California,* Riverside, California, 92502.

101

together with the stringent requirements imposed by government regulations has virtually eliminated all of these as available effective acaricides for the control of citrus red mite in important citrus growing districts in California. This dearth of effective acaricides has stimulated interest in the development of new acaricides.

The chemical industry has made or is capable of making hundreds of chemicals each year which are sufficiently toxic to mites to be potential acaricides. Although a high degree of toxicity is an essential property of an effective acaricide, other qualities often more difficult to evaluate are essential. Among these accessory qualities which require evaluation for any pesticide chemical are phytoxicity, penetration into plant tissue, further translocation and the ultimate sites of concentration in the plant, *in situ* degradation or other alteration products, and the influence of repeated applications on the development of resistance. Each of these factors requires complicated and prolonged study before making the initial selection of a compound for evaluation as a new miticide. Initial laboratory techniques which produce results that correlate with acaricide efficiency under field conditions rather than with merely the degree of toxicity are essential in order to prevent non-effective materials from being used in the necessarily elaborate and expensive field evaluations. Ebeling (1963) has shown that surface weathering degradation, penetration, translocation, and *in situ* stability and thus persistence all influence the duration of residues toxic to mites. Because of the operation of these factors on or in the plant tissue, there appears to be little correlation between the residues found on or in citrus fruit or leaves by chemical analysis and the residues which serve to control the pest (effective residues).

This difference between chemical and biological assay probably shows maximum expression when acaricides are applied to citrus because of the amount of citrus oil in leaves or in fruit rind. Although the cuticles of leaves contain wax, the amount varies with the plant, leaf age, and weather (Crafts and Foy 1962). Since tetranychid mites feed by extracting the content of the epidermal and palisade cells, an acaricide residue is effective only if the mites come in contact with the residues on or in the surface wax, or feed on cells into which the toxicant has penetrated. As many of the acaricides are lipoid-soluble, the major path of penetration is through the wax and epidermal cells and into the oil sacs (Crafts and Foy 1962, Linskens et al. 1965). It would be expected that a lipoid-soluble toxicant becomes rapidly concentrated in the oil. That this is the case is illustrated by radiotracer studies of Metcalf et al. (1954) as well as by the relative amounts of toxicants found in extracted oil in relation to that in the whole rind (reviewed by Gunther 1969). As citrus oil is toxic to mites it is likely that the "sacs" are avoided during mite feeding. Thus the acaricide may become unavailable to the mites either by contact or by feeding.

The density of citrus foliage makes it unusually difficult to distribute nonsystemic toxicants thoroughly enough to reach all the mites and their eggs on a citrus tree. Experience has indicated that the residues of acaricides which are effective in controlling citrus red mite effect a high mortality rate of at least one active mite stage for an interval sufficient for the eggs to hatch and for the susceptible active stage to contact the toxic residues. The duration of the residue available to the mites should, therefore, be closely correlated with the general effectiveness of the acaricide.

The amount of initial spray deposit and the persistence of the resulting residues on citrus fruits have been evaluated during the past 20 years for 14 acaricides previously known to control effectively the citrus red mite and other mites injurious to citrus in California. The primary purpose for chemically determining these residues was to provide information relative to the safe use of these acaricides on citrus fruit and thus in fruit products. The desirability of correlating the amount of toxicant and its rate of degradation, as determined by chemical analysis, with its biological activity against the citrus red mite was recognized. Therefore, when samples were harvested for chemical residue evaluation parallel samples were used to ascertain the toxicity of the deposits or residues to this mite.

The purpose of this paper is to present, review, and correlate the data obtained in these chemical and biological studies in order to provide not only basic information on the best use of these acaricides but to aid in evaluating dosages and residues of new compounds as potential acaricides.

II. Methods

Sprays were applied by conventional high pressure sprayers and trees from which residue samples were taken were sprayed from the center half of each tank of spray. Fruit samples for residue studies were harvested at pre-planned intervals after treatment (GUNTHER 1969). The samples for chemical residue evaluations consisted of 32 fruit/plot, one from each quadrant of each of eight trees, taken at shoulder height around the periphery of the trees. One additional fruit/tree was used for the bioassay evaluations, thus providing eight fruits/sample.

Methods of extracting and analyzing the residues for chemical assay varied with each acaricide (GUNTHER 1969). Bioassays were accomplished by providing an opportunity for about 50 mites to crawl onto each of the eight fruits. The mites were exposed to the residues for 48 hours, then dead and live mites were counted under 20X magnification.

III. Results

The amounts of deposits and/or the persistences of the resulting residues on and in the fruits following conventional spray applications are summarized below for each acaricide. Residue assays were made by both chemical and biological methods. The rates at which these residues became inactive to citrus red mite are also reported and correlated with the amounts of the acaricide found in the rind by the chemical analyses.

That biological activity of a toxic chemical to citrus red mite frequently has little correlation with the amount of residue present on or in citrus rind is pointed out; postulations are made as to the reasons for the relationship of or lack of correlation between the dosages applied and their resulting deposits as well as the differences in the persistence of the resulting residues, their effective toxicities, and the general effectiveness of the acaricides tested. The acaricides are presented according to the chronological order of their availability and are defined chemically according to Gunther *et al.* (1968).

a) Neotran

In the late 1940's Neotran [methane, bis-(*p*-chlorophenoxy)-] replaced DN-111 (the dicyclohexylamine salt of dinitro-*o*-cyclohexyl-phenol) as the major organic acaricide for control of the citrus red mite in California. It was observed that Neotran applications were less effective during the summer months than at other times of the year. This behavior prompted a study to ascertain the influence of temperature on the effective persistence of Neotran residues. At that time no satisfactory chemical method was available to analyze these residues.

The toxicities of Neotran deposits and residues were determined through bioassay at intervals following both a winter and a summer application (Jeppson *et al.* 1951). A summary of the results (Fig. 1A) shows that Neotran residues on navel oranges were effective for a longer period following a December application (line *a*) than during August (line *b*); also residues on fruit exposed under constant temperatures in the laboratory at 70° F. (line *c*) were effective longer than residues subjected to temperatures maintained at 88° F. (line *d*). One may conclude that the temperature to which Neotran residues are exposed largely determines the rate at which degradation of the compound to less toxic products occurs, thus influencing the availability of the residues to the mites.

The Neotran residues found following an application on the oranges of 0.8 lb./100 gallons on November 13, 1956, and another application on February 5, 1957, were sufficiently similar to be represented as a single line (Fig. 1B). The resulting deposits of 15 p.p.m. decreased to residues of seven p.p.m. in 40 days.

Applications from which residues for bioassay were obtained and those from which the residues were analyzed by chemical means were made during different years. Both were applied during the winter, however, and therefore the weathering rate should be approximately the same. The amounts of residue required to obtain mite mortalities under

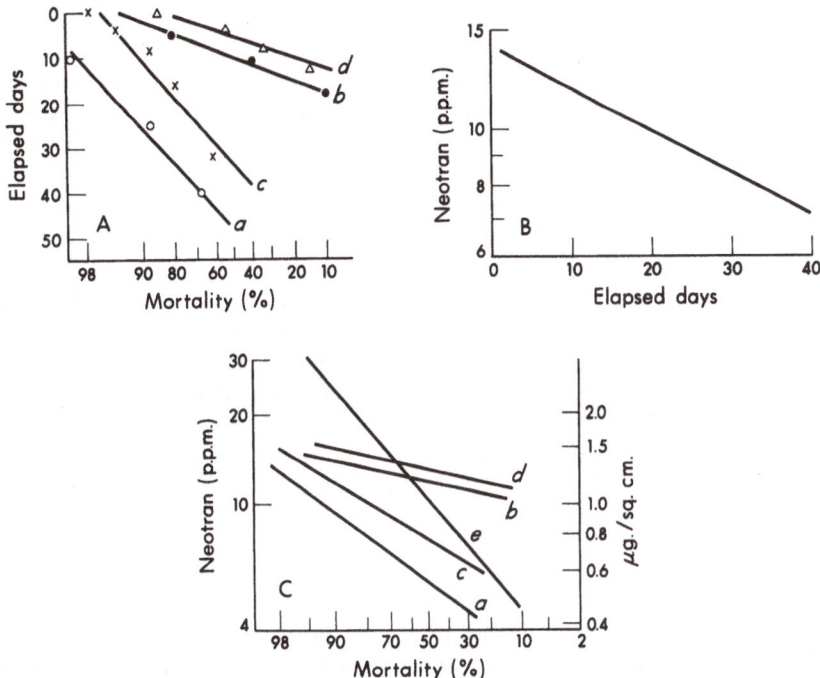

Fig. 1A. Mortality on navel oranges of citrus red mite exposed to Neotran residues under different temperature conditions following an application of 0.8 lb./100 gallons. Field: a = applied in December and b = applied in August. Laboratory: c = fruit held at 70°F. and d = fruit held at 88°F.

Fig. 1B. Rate Neotran residues on and in rind decrease with time as determined by laboratory bioassay

Fig. 1C. Relation of Neotran residues in p.p.m. to their toxicity to citrus red mite. Field: a = applied in December and b = applied in August. Laboratory: c = fruit held at 70°F., d = fruit held at 88°F., and e = residues in μg./sq. cm.

different conditions (Fig. 1C) were obtained by plotting the percent mortality found by bioassay in relation to the p.p.m. found by chemical analysis at the same interval after application. Such correlation from field applications on December 15 and August 27 are represented by lines a and b, respectively, and laboratory treatments held at 70° F. and 88° F. by lines c and d, respectively. METCALF (in JEPPSON 1951) determined the amount of Neotran on leaves in μg./sq. cm. required for toxicity to citrus red mite; these are represented by line e with the key on the right side of the graph. These results show that the effective residues are influenced by the temperatures prevailing during the weathering period.

GUNTHER and BLINN (1955) concluded, after studying the persistence and degradation of many pesticides, that "there is good agreement among half-life values for a given insecticide and a given substrate regardless of the dosage, time of year, geographical location of the experiment, and possibly minor type of formulations used, and varietal effects appear to be small." The results reported in Figure 1A show that the time of year applications were made (which reflects temperatures occurring during the weathering period) does influence the duration of effectiveness of these residues. One may conclude, therefore, that—at least with Neotran applied to citrus—there is poor correlation between the amount of residue present on and in the rind of citrus fruit and the toxicities of these residues to mites. For example, following the December 15, 1948, application of Neotran (Fig. 1C, line a), a residue of 5.5 p.p.m. resulted in 50 percent mortality, whereas it required 12 p.p.m. to effect the same mortality following the August 27 treatment (line b). The laboratory evaluations at 70° F. and 88° F. show similar relationships (Fig. 1C lines c vs. d). When mite mortality was plotted in relation to amount of Neotran residues expressed in μg./sq. cm. (Fig. 1C, line e) and in p.p.m., a 10-to-1 relationship was evident.

b) EPN

EPN (phosphonothioic acid, phenyl-O-ethyl O-p-nitrophenyl ester) was not used commercially for the control of citrus red mite in California; however, field experiments showed that spray concentrations of 0.25 lb./100 gallons resulted in effective control of this mite. The amount of deposit following applications of 0.5 and 0.25 lb./100 gallons to Valencia orange fruit and the rate at which the resulting residues weathered were ascertained (Fig. 2A).

The toxicities of the residues to citrus red mite were also evaluated from parallel fruit samples of each dosage. When mortalities were plotted in relation to the amount of residue in p.p.m. found by chemical analysis (Fig. 2B), the effective dosage range was from 7.0 to 0.7 p.p.m., and the LC-50 occurred at about one p.p.m. Although the initial deposits resulting from the 0.5 lb./100 gallons dosage were higher, the amount of residue remaining after about 43 days exposure was 0.95 p.p.m. for both dosages.

As field applications indicate that the 0.25 lb. dosage was as effective as the 0.5 lb. concentration (JEPPSON 1950), it appears that initial deposits of this acaricide above two p.p.m. on a whole-fruit basis (10 p.p.m. on a rind basis) are not essential.

The amount of EPN residue at a given interval after application appears to be closely correlated with mite toxicity. At an exposed period of 20 days following an application of 0.25 lb./100 gallons of spray, for example, there was about 1.4 p.p.m. found in the rind resulting in a mite mortality of 75 percent. At the higher dosage of 0.5 lb./100 gallons, 1.7 p.p.m. were required to produce the same mite mortality; however, this did not occur until 45 days after the application.

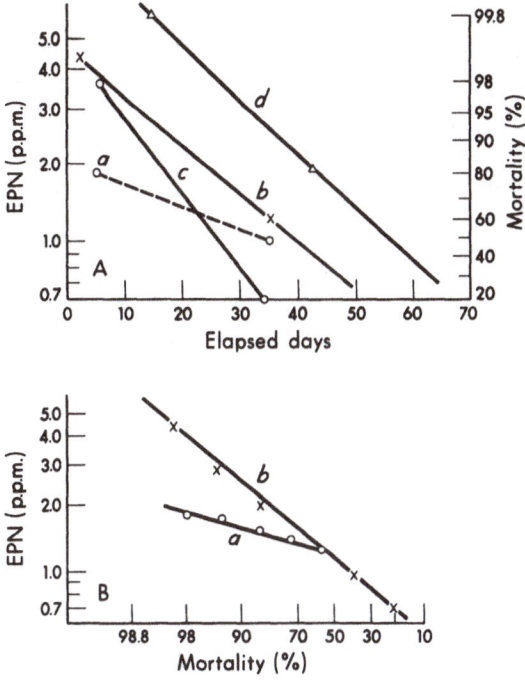

Fig. 2A. Decreases in actual and effective EPN residues on and in orange rind with time: a = p.p.m. found following application of 0.25 lb./100 gallons, b = p.p.m. found following application of 0.50 lb./100 gallons, c = percent mortality resulting from an application of 0.25 lb./100 gallons, and d = percent mortality resulting from an application of 0.50 lb./100 gallons

Fig. 2B. Relation of EPN residues in p.p.m. to toxicity to citrus red mite: a = 0.25 lb./100 gallons and b = 0.50 lb./100 gallons

c) Tetram

Tetram [O,O-diethyl S-2(diethylamino)ethyl phosphorothiolate] is an effective acaricide which was not used commercially because of its high contact mammalian toxicity. However, deposits and residues of this systemic toxicant were studied during 1955 by means of an acetylcholine method (GAGE 1961). The results of these analyses have not been published previously.

The amount of deposit resulting from an application of 40 g./100 gallons of spray and the rates of degradation and persistence of the resulting residues on navel oranges are graphically presented in relation to the number of days of exposure to the degrading factors in Figure 3A. Data on surface residues (line a) were obtained by washing the fruit and analyzing for the amount of Tetram in the wash water. The washed fruit was then peeled, and the rind was analyzed (line b). The total deposit and residue persistence measurements were obtained

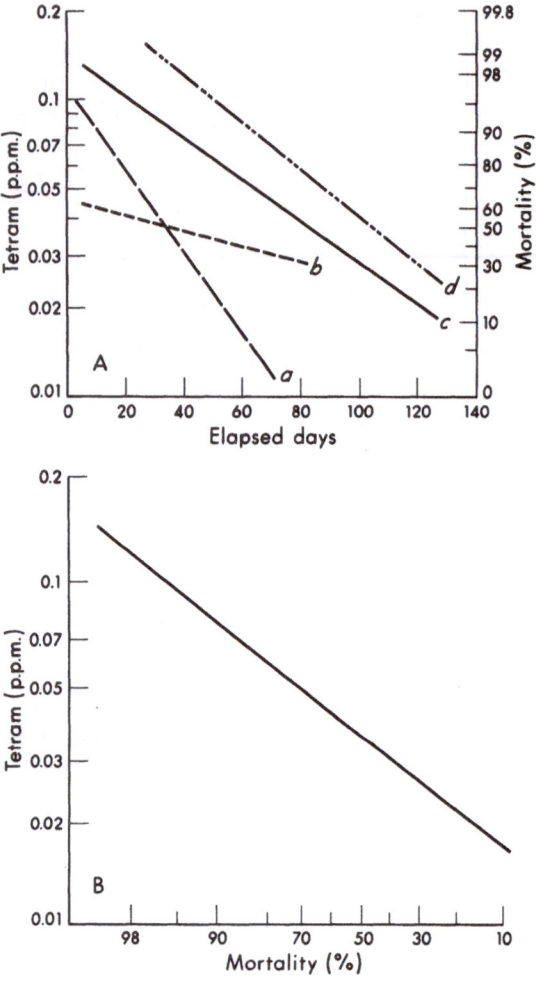

Fig. 3A. Decreases in actual and effective Tetram residues on and in whole navel oranges with time following an application of 40 g./100 gallons: *a* = surface residues, *b* = washed fruit-residues in rind, *c* = total residue, and *d* = percent mite mortality—unwashed fruit

Fig. 3B. Rate Tetram residues decrease in toxicity to citrus red mite

by extracting and analyzing the rind from unwashed fruit samples (line *c*).

The percentage mortality to adult female citrus red mites obtained at each sampling interval after application was indicated on a probit scale in relation to the elapsed days from treatment; the resulting correlation line is indicated in line *d*.

The results show that Tetram residues are relatively persistent on and in orange fruits. Surface residues decreased more rapidly (within 80 days) because of loss through physical weathering, by chemical degradation, and by absorption into and through the wax surface and cuticle into the fruit tissue (EBELING 1963, GUNTHER 1969).

Bioassays (line d) of Tetram deposits and residues indicate that not only are the Tetram deposits highly toxic, but the resulting residues are persistent and sufficiently available to mites to achieve a high degree of toxicity for a relatively long interval after application, with the LC-50 occurring at about 105 days. As the surface residues had decreased to a relatively nontoxic level in fewer than 80 days, it is evident that penetrated residues were available and toxic to the mites. As Tetram is water soluble it is likely that penetrated residues remain in plant cells, and are thus available to the mites if the toxicant remains unchanged in the cells from which the mites obtain their food instead of concentrating in the oil sacs, as happens with many acaricides. It would be expected, therefore, that a close correlation should exist between the percent mortality resulting from residue exposure to the amount found by chemical assay; that this is the case is indicated by the similarity in slopes of total residue line c and mortality line d in Figure 3A.

By plotting the amount of residue in p.p.m. found at a given interval (days) after treatment as points on the x-axis, having a log-scale in relation to percent mite mortality obtained at the same time as points on a probit scale on the y-axis, a straight-line relationship is obtained from which one readily can correlate residues in p.p.m. with percent mortality of mites. The effective residues ranged from 0.15 to 0.02 p.p.m., with the LR-50 occurring at 0.037 p.p.m.

As the lethal residue line and that of the actual residues decreased with time at relatively the same rate, any movement of Tetram from surface to penetrated residues appears to have a minimum influence on their availability to citrus red mite.

d) Ovex

Ovex (benzenesulfonic acid, p-chloro-, p-chlorophenyl ester) was used for a number of years for control of tetranychid mites in California and other citrus areas. Its use was terminated only where mite eggs and nymphs had developed resistance. The amount of ovex deposit and the rate resulting residues persisted on and in lemon rind following an application of one lb. of 50 percent wettable powder formulation/100 gallons were evaluated by GUNTHER and JEPPSON (1954) (Fig. 4A). They arbitrarily segregated the residues by means of three "stripping" methods into those adhering to the wax layer or true surface residue obtained by removal with two percent detergent wash (line a), those embedded within or dissolved in the wax "layer" removable by quick laving with a suitable solvent (line b), and the residues which had penetrated below the wax "layer" and removable only by thorough equilibration of the finely ground rind with the sol-

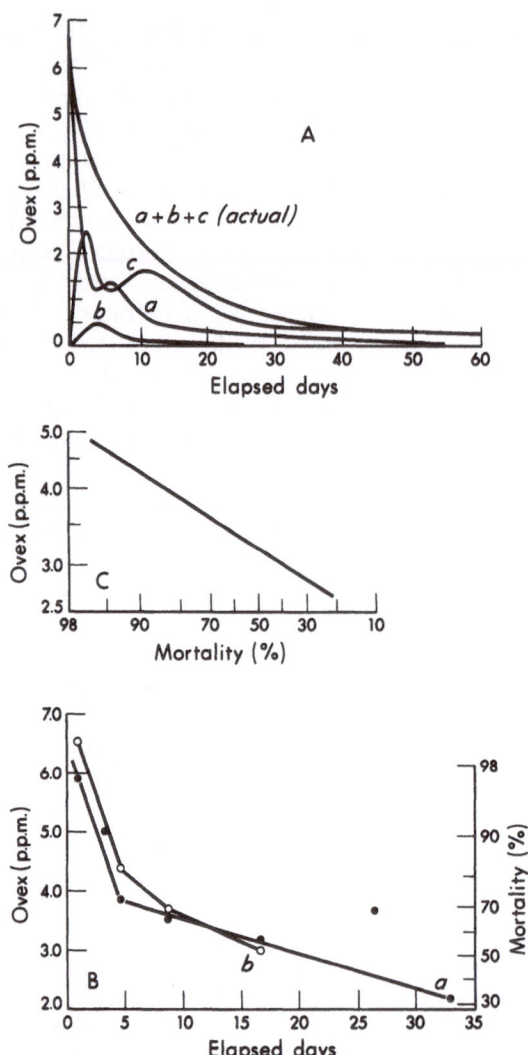

Fig. 4A. Penetration behavior of residues of ovex in lemon rind following an application of one lb./100 gallons. The $a + b + c$ (actual) curve is the smoothed curve summed from curves a, b, and c. (Reproduced after Gunther et al. 1969)

Fig. 4B. Decrease in actual and effective ovex residues with time: a = p.p.m. residue and b = percent mortality

Fig. 4C. Relation of ovex residues in p.p.m. to their toxicity to citrus red mite

vent (line c). It was found that within two to three days 75 percent of the acaricide had migrated from the surface into the cuticular waxes. The concentration within the wax "layer" soon reached an equilibrium, indicating that the acaricide traverses this layer in both directions. In the meantime, there was penetration into the rind reaching a maximum level within nine to 11 days. After 15 days all three types of residues had essentially stabilized in rates and direction of migration.

When mites were exposed to the residues on lemon fruit at each harvest interval, there was a rapid decrease in toxicity during the first five days at which time less than 50 percent of the exposed mites were killed by the residue (Fig. 4B). This behavior indicated that most of the toxicity of ovex residue to mite eggs and immature stages resulted from their exposure to the surface residues. When the mortalities were plotted against p.p.m. of the acaricide found by chemical evaluation (Fig. 4C) residues of 6.0 to 3.0 p.p.m. covered the toxicity range of ovex.

The major effectiveness of ovex residues, in contrast to those of Tetram residues, appears to be from surface residues. As this acaricide is primarily toxic to the egg and larval stages, the major exposure of the susceptible stages is by means of contact with the surface residues.

e) Aramite

Aramite [sulfurous acid, 2 (p-tert. butylphenoxy)-1-methylethyl-2-chloroethyl] became available during 1949 and was soon found to be highly toxic to the active stages of citrus red mite. Its effect was relatively slow, often requiring 96 hours to inactivate mites. It was used extensively as an acaricide on citrus until federal restrictions were imposed because extended animal feeding studies indicated that it had potential carcinogenic properties.

In order to detect residue migration and location in citrus rind, Aramite residues resulting from a 3.0 and 4.5 lb. actual/acre applica-

Table I. *Aramite residues on lemons* [a] *and navel oranges* [b] (GUNTHER et al. 1951)

| Citrus fruit | Date | | | Residue (p.p.m.) | | | |
	Applied	Sampled	Processed	Extra-cuticular	Cuticular	Sub-cuticular	Total
Lemons	9/6	9/6	9/9	0.20	0.10	0.60	0.90
	9/6	9/12	9/13	0.12	0.37	0.60	1.09
	10/30	10/30	10/31	0.09	0.09	0.17	0.35
	10/30	11/6	11/9	0.11	0.03	0.18	0.32
	10/30	11/13	11/21	0.02	0.02	0.14	0.18
Oranges	10/18	10/18	10/19	0.55	0.15	0.20	0.90
	10/18	11/6	11/8	0.20	0.12	0.24	0.56
	10/18	11/6	11/8	0.19	0.14	0.21	0.54

[a] Sprayed with 20 lb. of Aramite 15W/acre.
[b] Sprayed with 30 lb. of Aramite 15W/acre.

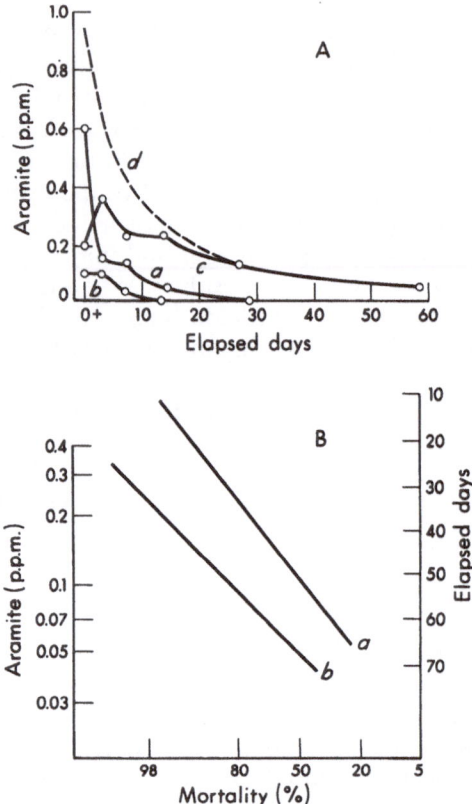

Fig. 5A. Penetration behavior of Aramite residues on whole lemons after spraying with two lb./100 gallons of 15 percent formulation: a = extracuticular, b = cuticular, c = subcuticular, and d = total residue. (Reproduced after Gunther 1969)

Fig. 5B. Rate Aramite residues decrease in toxicity to citrus red mite: a = mortality at indicated elapsed days and b = relation of Aramite residues in p.p.m. to their toxicity to citrus red mite

tion on lemons and oranges, respectively, were removed in steps as with ovex and reported by Gunther *et al.* (1951) as extracuticular, cuticular, and subcuticular residues (Table I). Also residues from sprays consisting of two lb. of a 15 percent formulation/100 gallons applied to lemon trees on October 30, 1950, were analyzed but not previously reported. These are graphically illustrated in Fig. 5A. The toxicities of Aramite residues to citrus red mite are shown in Fig. 5B, line a, and correlated with the total residue found by chemical analysis (line b).

About three-fourths of the initial deposit of Aramite becomes cuticular or subcuticular residue within three days (Fig. 5A). Maximum amount of cuticular residues occurred as soon as the samples could be extracted but had disappeared within 10 days. All residues had become subcuticular within 25 days of exposure.

Bioassay of these residues shows that even after 30 days weathering, the residues are toxic to over 80 percent of the citrus red mites. As all of the residues had become subcuticular prior to that time, it is evident that mites obtain this toxicant by feeding as well as by contacting surface residues. The total amount of residue in p.p.m. required for mite mortality (Fig. 5B, line *b*) is closely related to mite toxicity. About 90 percent mortality was obtained with residues below 0.15 p.p.m. which at that time were all subcuticular; thus it appears that most of the subcuticular residues are available and toxic to these mites as they feed on lemon fruit, suggesting that a minimum amount is concentrated in the oil sacs and that major amounts remain in the cells.

f) Chlorobenzilate

Chlorobenzilate (benzilic acid, 4,4'-dichloro-, ethyl ester), has not proven sufficiently toxic to the citrus red mite to be effective at practical dosages, but it is effective in the control of the citrus rust mite, *Phyllocoptruta oleivora* (Ashm.), and the citrus bud mite, *Aceria sheldoni* (Ewing) (JEPPSON *et al.* 1955). An application of four ounces of chlorobenzilate/100 gallons of spray resulted in an initial deposit of 9.5 p.p.m. on and in lemon rind and residues that persisted for more than 160 days (Fig. 6A). Initial deposits failed to kill 75 percent of the citrus red mite adult female mites (Fig. 6B); however, an application of 12 oz. actual/100 gallons (higher than practical dosage) resulted in an initial deposit of 19 p.p.m., an amount sufficient to effect a high initial mite mortality. Also, the residues remaining on the fruit were above nine to 10 p.p.m., the level which produced 75 percent mite mortality.

There was a good correlation between chlorobenzilate residues found by chemical analysis and their toxicities to citrus red mite as indicated by comparing the chemical residue and bioassay data. When these results are plotted on log-probit residue mortality scales, the points resulting from both dosages make a single residue-mortality line, with the LR-50 occurring at 7.5 p.p.m. This suggests that much of the chlorobenzilate residue remains on or in tissue available to citrus red mite by contact or feeding and, therefore, is not concentrated readily in the oil sacs which are avoided by the mites.

This characteristic of long residue persistence on the surface where the toxicant is available for direct contact or in the tissue where the mites feed provides an explanation as to why this acaricide is unusually effective against the citrus rust and bud mites. These mites are more susceptible to chlorobenzilate than is the citrus red mite, and acaricides effective against these eriophyid species must be toxic to the mites that move out of nitches that provide protection from even the most thorough kind of acaricide application.

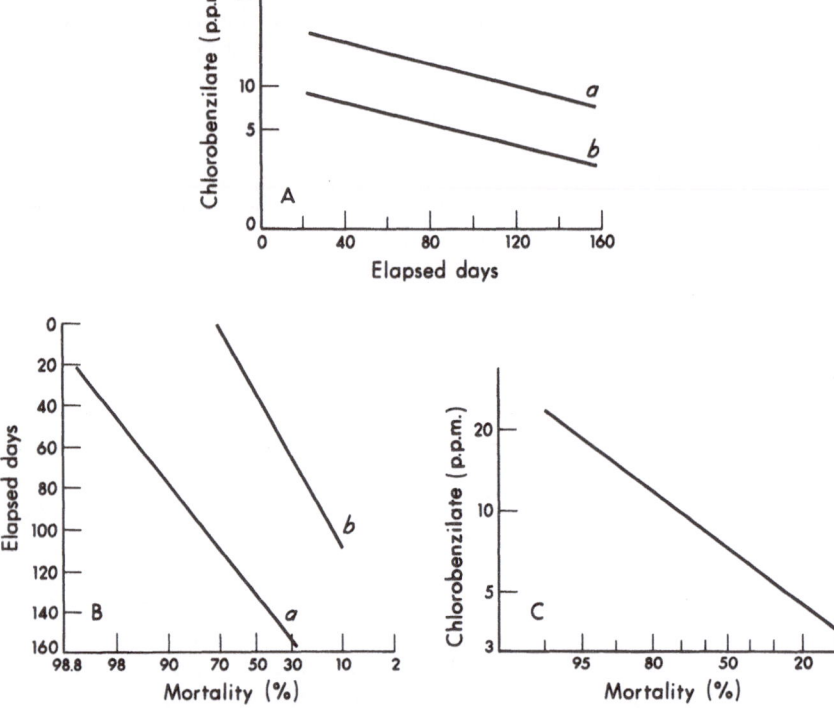

Fig. 6A. Rates chlorobenzilate residues on and in lemon rind decrease with time:
          a = ¾ lb./100 gallons of spray and b = ¼ lb./100 gallons of spray.
          (Reproduced after GUNTHER et al. 1969)
Fig. 6B. Rate chlorobenzilate residues decrease in toxicity to citrus red mite:
          a = ¾ lb./100 gallons dosage and b = ¼ lb./100 gallons dosage
Fig. 6C. Relation of chlorobenzilate residues in p.p.m. to their toxicity to citrus
          red mite

g) Carbophenothion (Trithion)

Carbophenothion {phosphorodithioic acid, S-[(p-chlorophenyl)thi-
omethyl]-, O,-O-diethyl ester} was found to be effective in controlling
citrus red mite in California prior to the development of resistance to
the organophosphorus compounds (JEPPSON et al. 1957). It has not
been used extensively because populations resistant to demeton were
highly cross-resistant to this acaricide (JEPPSON et al. 1958).
    The persistence of carbophenothion residues was evaluated by
chemical analysis and bioassay following applications involving three
concentrations of emulsifiable and wettable powder formulations to
both lemon and orange trees. The initial deposits and the weathering
and degradation of residues with time are graphically illustrated in
Figure 7A (GUNTHER et al. 1959). The initial deposit from equivalent

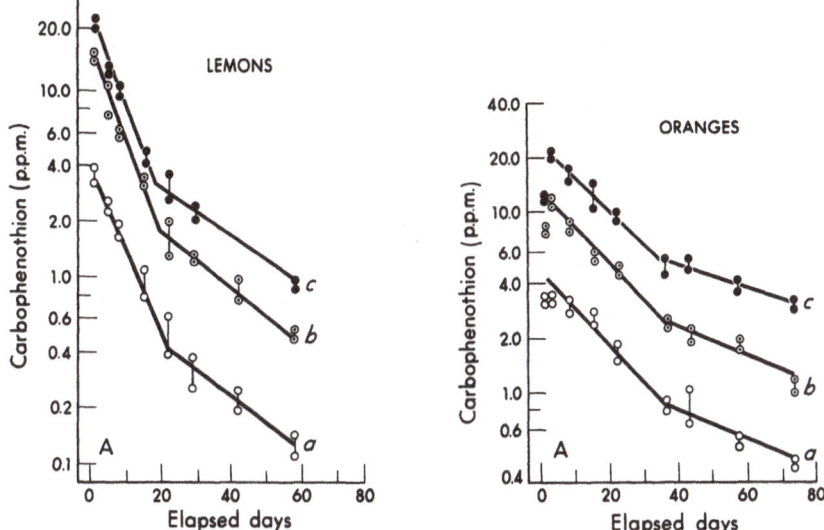

Fig. 7A. Rates carbophenothion residues on and in rind decrease with time following applications of a 25 percent WP formulation to lemons and
oranges (reported as 1b. actual compound/100 gallons of spray): $a = 1$,
$b = 3$, and $c = 6$. (Reproduced after GUNTHER et al. 1969)

dosages was about the same on oranges as on lemons; however, as
lemons were growing during the weathering period, the amount of residue decreased more rapidly. Residues from all dosages decreased
rapidly for 15 days on lemons and 35 days on navel oranges due to the
combination of degradation and weathering, after which the slower
decrease indicated loss by degradation within the rind.

The number of days carbophenothion residues were toxic to citrus
red mite varied with the dosage, the formulation, and the citrus
variety. Residues were highly toxic throughout the first 20 days of exposure on lemons and 35 days on oranges (Fig. 7B). The period during which residues of a given formulation were effective was closely related to the dosage applied. An emulsifiable concentrate (EC) formulation applied at three oz./100 gallons resulted in equivalent deposit, but
longer actual and effective persistence than the wettable powder (WP)
formulation applied at the same amount of toxicant/100 gallons.

By correlating residues in p.p.m. at different elapsed days in relation to percent mortality at the same elapsed time interval (Fig. 7C),
it is evident that the lower the dosage, the more effective the mortality/
amount of toxicant. The three oz./100 gallons dosage of emulsive concentrate was less effective at equivalent amount of residue than the
same amount of residue from the application of the wettable powder.

These findings show that formulations as well as dosage have an
important influence on both the actual and the effective persistence of

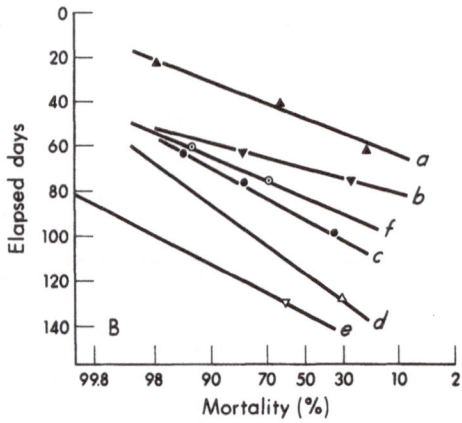

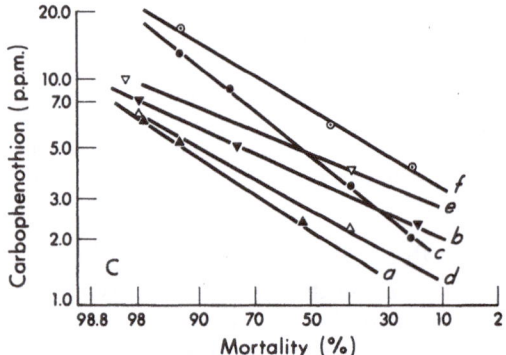

Fig. 7B. Effect of dosage, formulation, and host on the rate carbophenothion residues decrease in toxicity to citrus red mite (applied at indicated oz./ 100 gallons of spray): $a = 1$ to lemons as WP, $b = 3$ to lemons as WP, $c = 6$ to lemons as WP, $d = 1$ to navel oranges as WP, $e = 3$ to navel oranges as WP, and $f = 3$ to lemons as EC

Fig. 7C. Relation of carbophenothion residues in p.p.m. to their toxicity to citrus red mite (applications as reported in Fig. 7B)

this acaricide. Also, the actual residues present and their effectiveness are not always closely correlated, suggesting that at least part of the residues that penetrate into the rind become unavailable to mites.

h) Dioxathion (Delnav)

Dioxathion (p-dioxane-2, 3-diyl ethyl phosphorodithioate) was found to be an effective acaricide in controlling citrus red mites in California before resistance to the systemic acaricide demeton {phosphorothioic acid, O,O-diethyl-O-[2(ethylthio)ethyl] ester mixed with O,O-diethyl-S-[(2-ethylthiol)ethyl]ester} had developed. As cross resistance to dioxathion was less than to carbophenothion, and because dioxathion

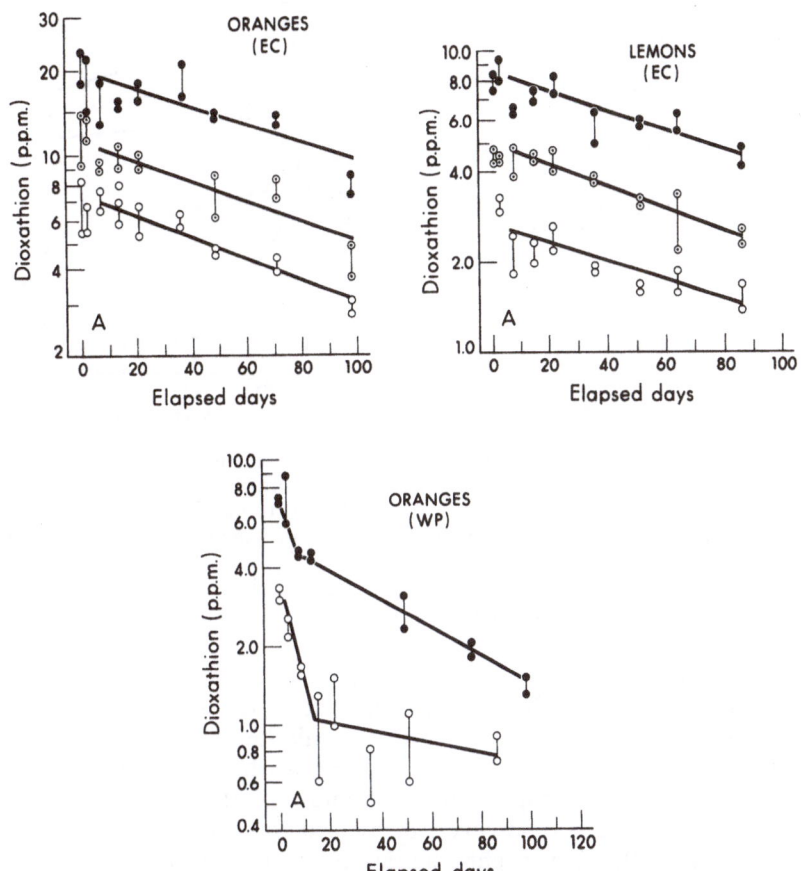

Fig. 8A. Rates dioxathion residues on and in rind decrease with time following applications of 25 percent formulations to navel oranges or lemons. Curves indicate p.p.m. or percent mortality resulting from applications of formulations and concentration/100 gallons indicated below (reported as oz./100 gallons of spray): $a = 3$, $b = 6$, $c = 12$, and $d = 24$

is also effective in controlling citrus thrips, *Scirtothrips citri* (Moulton), it has had more usage as a pesticide on citrus in California.

The dioxathion deposits and the relative persistence of the resulting residues were evaluated following three application dosages of a four lb./gallon emulsifiable concentrate (EC) formulation on lemons and on oranges and one dosage of the EC and a 25 percent wettable powder (WP) formulation on lemons (Fig. 8A) (GUNTHER *et al.* 1958).

The deposits resulting from applications of the EC formulation were considerably higher on oranges than on lemons; however, the initial deposit of the WP formulation was higher on lemons than on oranges (Fig. 8B). This fact suggests that the two fruit surfaces have different

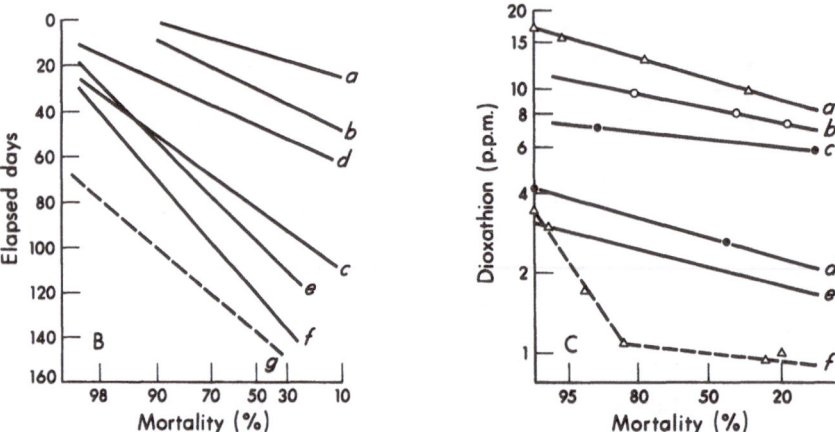

Fig. 8B. Effect of dosage and host on the rate dioxathion residues decrease in toxicity to citrus red mite (applied as EC formulations at indicated oz./ 100 gallons): $a = 3$ to navel oranges, $b = 6$ to navel oranges, $c = 12$ to navel oranges, $d = 6$ to lemons, $e = 12$ to lemons, $f = 24$ to lemons, and $g = 8$ to Valencia oranges

Fig. 8C. Relation of dioxathion residues in p.p.m. to their toxicity to citrus red mites (applications as reported in Fig. 8B)

requirements for spray deposition and each formulation was more suited to one than the other fruit for depositing the acaricide.

Deposits resulting from the EC formulations apparently moved rapidly into the rind allowing little surface weathering as indicated by a straight-line relationship between amount of deposit and elapsed days. The WP residues, however, decreased rapidly the first 10 days, then followed the consistent degradation pattern, indicating that weathering was the major cause of loss of acaricide during the first 10 days.

Effective toxicity evaluations of dioxathion residues (Fig. 8B) indicate that the toxicity period was longest on Valencia oranges and shortest on navel oranges. Thus, the effective residue period increased as dosages were increased from three to 24 oz./100 gallons. The LR-50 on lemons ranged from 45 days when six oz. were applied to 118 days following the application of 24 oz./100 gallons. The LR-50 on navel oranges from 13 days, following the three oz./100 gallons application, to 80 days after a 12 oz./100 gallons dosage.

By correlating the p.p.m. residue found on the fruit with mite mortality (Fig. 8C), it is apparent that much less dioxathion residue was required on lemons to effect equivalent mite mortality than on navel oranges. Also the toxicity of dioxathion residues/p.p.m. decreased as the original dosage and the amount of residue found were increased; however, the wettable powder was more effective than the emulsifiable formulation/unit of residue found in the rind. All these data indicate that residues of this acaricide migrate into portions of the rind which

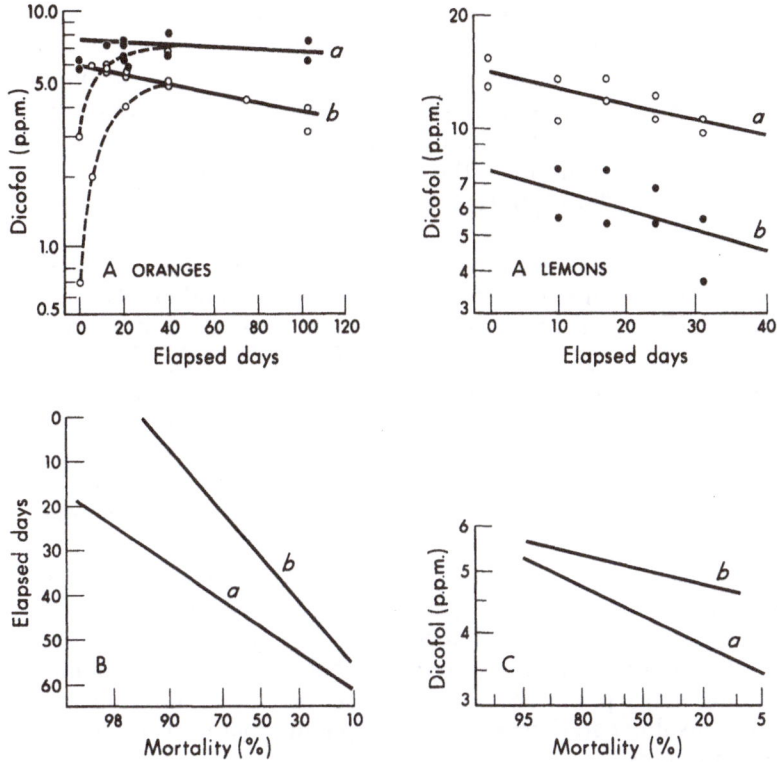

Fig. 9A. Rates dicofol residues on and in rind decrease with time following applications of 25 percent formulations at 0.4 lb. actual/100 gallons to Valencia oranges or lemons: *a* = EC, broken line = washed fruit and *b* =
WP, broken line = washed fruit (Reproduced after GUNTHER *et al.* 1969)

Fig. 9B. Rate dicofol residues decrease in toxicity to citrus red mite: *a* = lemons
and *b* = Valencia oranges

Fig. 9C. Relation of dicofol residues in p.p.m. to their toxicity to citrus red mite:
a = lemons and *b* = Valencia oranges

are not available to the mites either by contact or by feeding or else
the acaricide is not toxic as a stomach poison, *i.e.*, it is toxic only by
contact.

i) Dicofol (Kelthane)

Dicofol [benzyhydrol, 4,4'-dichloro-*a*-(trichloromethyl)-] is a general acaricide which has been used extensively for the control of tetranychid, tenuipalpid, and other mite pests. It has been used since 1957 for
control of these mites on citrus in California.

More extensive chemical residue evaluations were made with this
acaricide than any discussed thus far; these include analysis by two
methods on two citrus varieties (lemons and oranges) and two types

of formulations (Gunther *et al.* 1957). Residue studies on Valencia oranges included a series in which the fruits were washed in a detergent solution prior to being peeled and processed so as to obtain information as to the ease of removal of deposits and extrasurface residues. These authors conclude that the residue persistence of dicofol in citrus rind is unusually long, the half-life ranging from 170 to 350 days on oranges and 120 to 150 days on lemons, depending upon the analytical method. Data on washed and unwashed fruit show that dicofol penetrates into the oily and waxy poritons of the rind rather slowly. Thus, residues from both wettable powder (WP) and emulsifiable concentrate (EC) applications harvested prior to 40 days after treatment may be reduced by the washing procedure (Fig. 9A). These residue studies indicate that degradation, physical loss by vaporization, or dislodgement are negligible once residues enter the rind. The differences in the deposits resulting from the two formulations can be attributed to the difference in their depositing characteristics. Fruit growth accounts for the apparent differences in residue persistence between lemons and oranges.

Initial deposits of six p.p.m. produced by applications of the WP formulation at four oz. actual chemical/100 gallons of spray to Valencia oranges resulted in 98 percent mortality of mites (Fig. 9B), but when the total residue on oranges had decreased only one p.p.m., the remaining five p.p.m. became toxic to 50 percent of the mites; by the time the residue had decreased to four p.p.m., there was no measurable toxicity to mites.

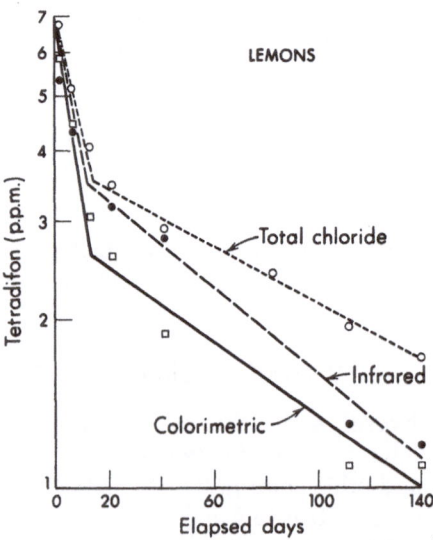

Fig. 10. Rate tetradifon residues decrease with time following applications of one lb./100 gallons of a 25 percent WP formulation when analyzed by different methods. (Reproduced after Gunther 1969)

The results of these chemical and biological assays suggest that either dicofol is non-toxic to the citrus red mite when taken up through the digestive system, or the residues are readily absorbed into the oil sacs in the fruit or other plant tissue not reached by the mites.

The residue on oranges resulted in less mite mortality than the equivalent amount of residue on lemons (Fig. 9B, compare lines a and b). As the dicofol was applied to oranges in May and to lemons during January, the process of adsorption and translocation to the oil sacs of the fruit rind was probably more rapid during the warmer conditions to which the residues on oranges were exposed.

j) Tetradifon (Tedion)

The ovicide-larvicide tetradifon (sulfone, p-chlorophenyl-2,4,5-tri-chlorophenyl-) was effectively used for control of tetranychid mites on citrus in California until the mites developed resistance to its application.

Residue studies by several analytical procedures indicated that there was little if any loss of residual tetradifon on Valencia oranges or lemons, even after 100 days (CASSIL and FULLER 1958, GUNTHER 1969) (Fig. 10); also, detergent washing two days after treatment with one lb. of a 50 percent WP formulation removed only minor quantities of the

Table II. *Residues of tetradifon on and in rind of field-treated lemons before and after commercial washing operations* (BLINN et al. 1959)

Replicate	Residue (p.p.m.)		Residue removed (%)
	Unwashed	Washed	
1	1.25	0.46	63.2
2	1.53	0.67	56.2
Average	1.39	0.57	59.0

residual acaricide. BLINN et al. (1959) found that commercial washing operations which remove 20 to 25 percent of the wax also removed 60 percent of the tetradifon from lemon fruit (Table II). In addition, by using selective solvents BLINN et al. removed increasing amounts of the citrus wax and found that about 90 percent of the tetradifon residue was located within the waxy cuticle of the rind, and about 55 percent of the residue was coextracted with only two to four percent of the wax (Table III). Since the residue was not merely adhering to the surface, as shown by its resistance to washing, it may be concluded that a portion of a persisting tetradifon residue was embedded or dissolved in the outer portion of the waxy cuticle.

The toxicity of the residues to mites was not evaluated but field control results indicate that residues are effective for several weeks; thus, the larvae and eggs are killed by this acaricide even after it has

Table III. *Results of selective-solvent laving of Valencia oranges to remove tetradifon residues* (BLINN *et al.* 1959)

Solvent	Days after treatment	Volume solvent/fruit (ml.)	Residue (p.p.m.) in Laving	Residue (p.p.m.) in Rind	Percent Tetradifon in laving	Percent Wax and extractives in laving [a]
Chloroform	23	40	1.30	0.20	87	100 [b]
	38	40	1.21	0.18	87	100 [b]
n-Hexane	23	70	1.39	0.39	78	22 [c]
	38	40	1.16	0.51	70	20 [c]
Acetonitrile	23	20	0.74	0.61	55	2 [d]
	38	40	1.12	0.84	57	4 [d]

[a] Arbitrarily considering the chloroform laving as 100 percent.
[b] Hard, yellow wax.
[c] Hard, colorless wax.
[d] Soft, colorless wax.

been absorbed into the wax layer. The observation that rains occurring shortly after applications of tetradifon wettable powder have not adversely influenced control supports these conclusions.

k) OW-9 Compounds

Compounds OW-9 [sulfurous acid, 2-(*p-tert.*-butylphenoxy)-1-methylethyl-1-methylethyl-2-chloroethyl ester and the 1-methylethoxy analog] were studied as a potential acaricide for control of the citrus red mite and found to be effective. Residue studies were made in preparation for partitioning for registration, but completion of the animal feeding studies showed that the compound possessed carcinogenic properties and, therefore, it was never used as an acaricide.

Quantitative persistence comparisons of residues from several dosages on and in the rind of Valencia oranges and on and in foliage are reported by GUNTHER (1969). He found that the half-lives on and in leaf tissues are consistently about half those on and in rind tissue, indicating less "protection" on the leaf residue (Fig. 11A). Washing removed very little of the initial deposits from either substrate demonstrating very rapid penetration of this particular compound. On an area basis the leaves accepted initial deposits equivalent to those accepted by the fruits yet residues persisted longer on or in the fruit than on and in the leaves. On a weight basis (p.p.m.) the leaves accepted and retained apparently immensely greater deposits than the fruits, *i.e.*, the ratio for the leaves of p.p.m. to µg./sq. cm. was consistently five-to-one. It is apparent from the data presented that residues in p.p.m. on and in citrus foliage cannot be directly compared with residues in p.p.m. on and in citrus fruits without a large interpolation factor.

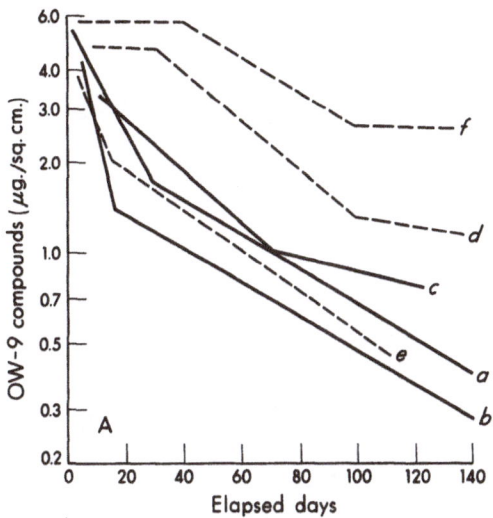

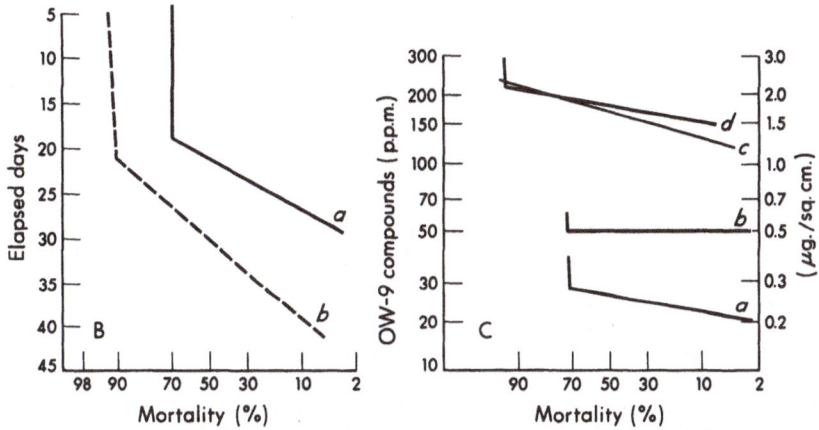

Fig. 11A. Rates OW-9 compounds residues on Valencia orange trees decrease with time after the following treatments: a = 5 oz./100 gallons on unwashed leaves, b = 5 oz./100 gallons on leaves washed after treatment, c = 9 oz./100 gallons on unwashed leaves, d = 5 oz./100 gallons on unwashed whole fruit, e = 5 oz./100 gallons on whole fruit washed after treatment, and f = 9 oz./100 gallons on unwashed whole fruit

Fig. 11B. Rate OW-9 compounds residues decrease in toxicity to citrus red mite with time: a = on fruit and b = on leaves

Fig. 11C. Relation of OW-9 compounds residues on leaves and fruit in p.p.m. and μg./sq.cm. with toxicity to citrus red mite: a = fruit in p.p.m., b = fruit in μg./sq.cm., c = leaves in p.p.m., and d = leaves in μg./sq.cm.

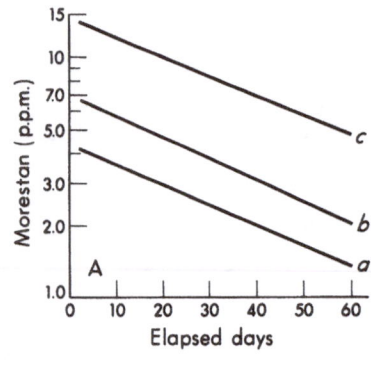

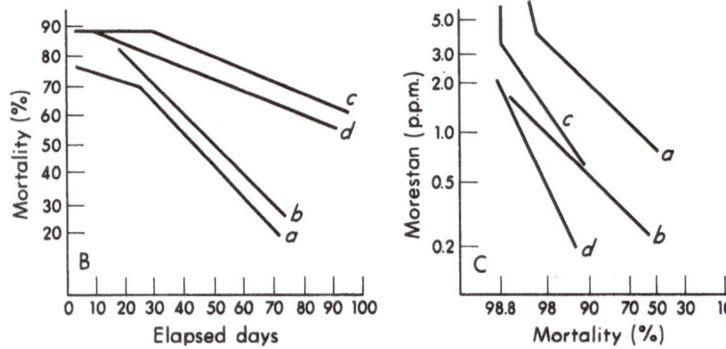

Fig. 12A. Rates Morestan residues on and in the rind decrease with time follow-
           ing 0.5 and 1.0 lb./100 gallons dosages to lemons or Valencia oranges:
           a = 0.5 lb. on lemons, b = 0.5 lb. on oranges, and c = 1.0 lb. on
           oranges

Fig. 12B. Rates Morestan residues decrease in their toxicity to citrus red mite
           or the percent that leave the fruit (loss): a = mortality on oranges,
           b = mortality on lemons, c = loss from oranges, and d = loss from
           lemons

Fig. 12C. Relation of Morestan residues in p.p.m. to their toxicities to citrus red
           mite or to the percent that leave the fruit (loss): a = mortality on
           oranges, b = mortality on lemons, c = loss from oranges, and d = loss
           from lemons. (Reproduced after Jeppson et al. 1967)

At each sampling interval part of the fruit and part of the leaves
were held to determine the toxicity of the residues to adult female
citrus red mite. Mortalities of citrus red mites on orange fruit, which
resulted from deposits and residues of the five, seven, nine, and 12 oz./
100 gallons dosages, did not increase as the dosage was increased. Initial deposits killed only 70 percent of the mites on fruit and 92 percent
on leaves (Fig. 11B). The toxicity to mites remained the same on both
leaves and fruit through 20 days, after which the effectiveness of the
residues decreased in relation to elapsed time.

When amount of residue (p.p.m.) on fruit and leaves at each interval of time is correlated with the mortalities obtained at the same time interval, it becomes evident that there is little correlation between p.p.m. or μg./sq. cm. residue found and toxicity to mites (Fig. 11C). The OW-9 compounds must rapidly migrate to parts of the rind not contacted by the mites. Either the residues migrate into the oil sacs or the residues in the rind are not toxic to mites as a stomach poison. As this phenomenon occurs on leaves as well as fruit, it is unlikely that the residues migrate into cells beyond which the mites normally feed.

l) Morestan

Morestan (carbonic acid, dithio-, cyclic S,S-ester with 6-methyl-2,3-quinoxalinedithiol) has been found to effectively control citrus red mite (JEPPSON et al. 1967). Chemical residue analyses indicate that initial deposits were about four to 14 p.p.m. with a half-life of 34 to 37 days for both lemons and oranges (Fig. 12A).

Observation on the response of citrus red mite to Morestan revealed that the mites strongly tend to leave the fruit bearing residues of this acaricide. It was, therefore, necessary to determine the percentage of mites that leave the fruit as well as the mortality of the remaining mites. Those leaving were reported as percent loss even though it was known that most of those migrating from the fruit eventually die. No doubt mites more tolerant to the chemical remained; therefore, at the same time interval after treatment there was a higher percentage loss than there was mortality of those remaining (Fig. 12B).

By plotting the amount of residue in relation to mite mortality the relationship of loss to mortality is more readily seen (Fig. 12C). Equivalent amounts of residue resulted in higher mite mortalities on lemons than on oranges. Residues above 3.5 p.p.m. on oranges and 2.0 p.p.m. on lemons did not increase the percent mite mortality or increase the migration. As the amount of residue decreased below 3.5 or 2.0 p.p.m., the percent mortality and loss by migration decreased in linear proportion to the amount of residue present. Residues of this acaricide are readily washed off fruit by rains or laboratory washing; this, together with the close relationship of residues found to mite mortality, indicates that Morestan residues largely remain on the fruit surface rather than penetrate into the wax or the rind.

m) Omite

Omite [sulfurous acid, 2-(p-tert.-butylphenoxy)cyclohexyl 2-propynyl ester] has been found to be effective in controlling citrus red mite in California (JEPPSON et al. 1969). Evaluations of residues on Valencia oranges were made by two methods (Fig. 13A) (GUNTHER 1969). Even though an emulsifiable concentrate formulation was applied, weathering proceeded at a rapid rate during the first 20 days. This was followed

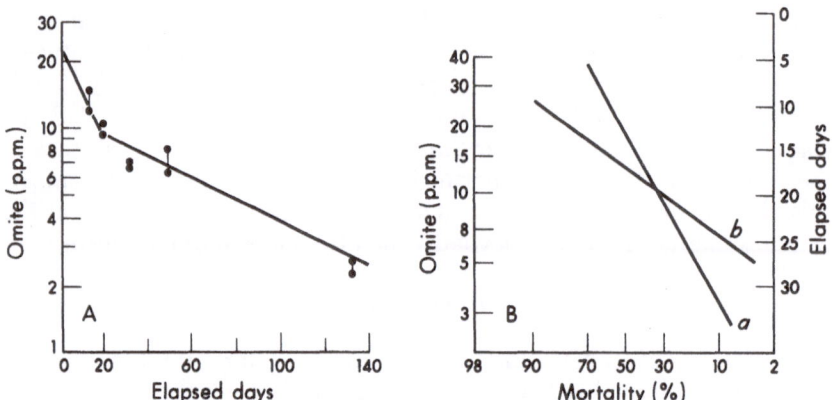

Fig. 13A. Rate Omite residues decrease with time following a 57 percent EC formulation at 0.5 lb. actual/100 gallons to Valencia oranges. (Reproduced after GUNTHER et al. 1969)

Fig. 13B. Relation of Omite residues in p.p.m. and time of residue exposure to their toxicities to citrus red mite. Application dosage was 5 oz./100 gallons: a = decrease in residues with time and b = toxicity in p.p.m.

by a less rapid rate attributable to the degradation processes within the fruit rind.

Toxicity of the residues by the described bioassay technique indicated that a low percentage of the mites is killed after weathering 20 days (Fig. 13B, line a). By correlating mortalities with the amount of residue present in p.p.m. (Fig. 13B, line b), it is also evident that mortality of the mites is largely dependent on surface residues. Only 30 percent of the mites was killed 20 days after treatment when chemical assay indicated residues of 10 p.p.m.

Migration of Omite residues, therefore, resembles that of dicofol, dioxathion, and OW-9 compounds, namely, the residues require a number of days to completely migrate into the rind and there concentration occurs in the oil sacs where the residues are unavailable to the mites.

n) Azodrin

Azodrin (phosphoric acid, dimethyl ester, ester with cis-3-hydroxy-N-methyl crotonamide) has recently been developed as an insecticide-acaricide and has been found to be effective for control of the citrus red mite on Valencia orange; however, by the third application to a field population the mites had developed a 12-fold (LC-50) resistance to this acaricide (Fig. 14).

The amounts of deposit and the rates of weathering and degradation of the residues were evaluated following a boom application at 2,000 gallons/acre containing 2.5 lb. of Azodrin/acre, and another application by air-blast type sprayer applied at 1.0 lb. in 50 gallons of

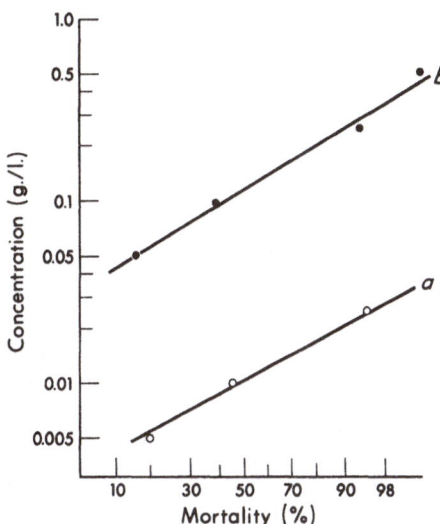

Fig. 14. Change in the susceptibility of citrus red mites to Azodrin as a result of repeated treatments: a = susceptible strain and b = after the third application

spray/acre (GUNTHER and JEPPSON 1969). The amount of residue found by chemical analysis following each application by analyzing the rind is graphically illustrated in Fig. 15A. When mites were exposed to the residues resulting from the higher dosage (Fig. 15A, line b) the mortalities that occurred are shown in Fig. 15B. At 12-, 26-, and 54-day sampling intervals the sample size was doubled and one-half the fruit was washed before the residues were extracted for analysis. At each sampling interval the residues found on the washed and unwashed fruit were the same, indicating that this compound rapidly and completely disappears from the surface of the fruit, *i.e.*, it penetrates.

Citrus red mites are normally highly susceptible to Azodrin when exposed directly to spray and to initial deposits (see Fig. 14, line a); therefore, one would expect field residues also to be highly toxic. However, residues of two p.p.m. were required to obtain 50 percent mortality (LR-50 in Fig. 15B), which indicates that mite toxicity results from residues in the fruit cells rather than from surface or contact-type residues. As mortality did not decrease more rapidly than the amount of residue, it appears that this acaricide does not migrate into the oil sacs but remains in the cellular tissue. When residues found and mite mortalities obtained are plotted on log-residue-probit scales, a straight-line relationship occurs (Fig. 15B) indicating that the rate of chemical degradation and the decrease in toxicity proceed at about the same rate. As mites feed on the epidermal and palisade cells and avoid the oil sacs, the total data indicate that this acaricide must remain in cell tissue

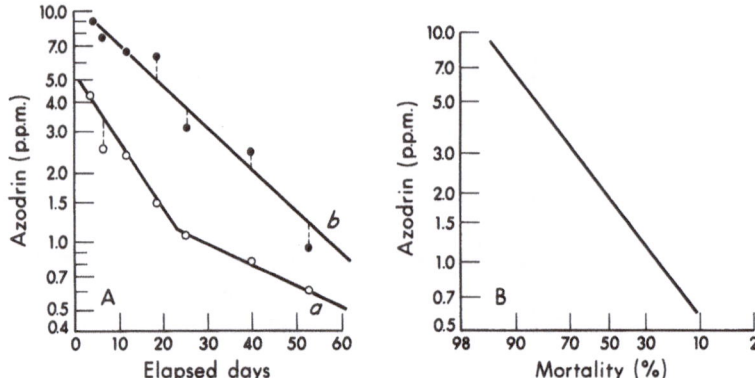

Fig. 15A. Rates Azodrin residues on and in Valencia orange rind decrease with time and toxicity to citrus red mite: a = applied by air blast equipment at 1.0 lb./acre in 50 gallons of spray and b = applied by boom equipment at 2.5 lb./acre in 2,000 gallons of spray

Fig. 15B. Relation of Azodrin residues in p.p.m. to their toxicity to citrus red mite on Valencia oranges

rather than migrate to oil sacs, and thus the decrease in mortality with time is strongly correlated with degradation of the chemical.

IV. Discussion of results

It is evident that there is often little correlation between the amount of residue found on and in citrus rind by chemical analysis and the effectiveness of the residues against citrus red mite because the effectiveness is dependent on rate of absorption, penetration, and the ultimate deposition of residues in citrus rind.

Mites may become exposed to toxic residues on citrus fruits and leaves by contact with surface residues or by feeding on the residues which have penetrated through the cuticle into the epidermal and palisade cells. Citrus rind contains many oil sacs where some acaricides appear to concentrate. For example, Metcalf et al. (1954) showed by radiotracer techniques that the thiono-isomer-metabolite of the insecticide demeton rapidly becomes located in the oil glands of lemon leaves, whereas the thiol-isomer-metabolite is much more generally distributed throughout the leaf tissue and they concluded that this distribution was due to the greater lipoid solubility of the thiono-isomer-metabolite. Also, in numerous publications, Gunther and co-workers (see Gunther 1969) have concluded that most nonsystemic insecticides and acaricide residues are ultimately distributed preferentially in the citrus oils after application to fruit near maturity.

Deposits of such acaricides as EPN, Tetram, Azodrin, and Aramite, however, appear to migrate rapidly into the aqueous fruit cells where the toxicant remains available to mites by ingestion. On the other hand, Morestan appears to remain on the surface where the residues are available to the mites by contact as they move about. In both such cases there is good correlation between the amount of residues found on and in the rind and their effectiveness.

Tetradifon and possibly some of the Neotran residues rapidly migrate into and remain in the surface waxes and thus are not readily removed by water washing yet appear to be available to the eggs and larvae of citrus red mites. As mite eggs deposited on tetradifon-treated fruit are killed by this acaricide, there must be movement of such compounds from the fruit to the eggs and larvae. Although data are not available on the movement of chlorobenzilate residues on citrus rind, their effective duration, their toxicity to eriophyid mites for an extended period even following rains, and the close correlation of the residues to their effectiveness indicates that much of the residue remains in the surface waxes.

Residue migration studies with ovex, Aramite, carbophenothion, and dicofol indicate that movement into the rind is a slow process enduring for several weeks. Residues of such acaricides decrease rapidly at first due to loss by weathering, but the rates of degradation, once residues are in the rind, are much slower. As these acaricides are highly toxic to mites only during the "weathering period" [*i.e.*, the so-called degradation period of GUNTHER and BLINN (1955)], it appears that each of these materials slowly migrates through the wax and the rind cells and finally is deposited in the oil sacs, thus becoming unavailable to mites either by contact or by feeding. Even though the amount of residue had decreased relatively little by the time mites were exposed to the fruit surface, the remaining residues were no longer toxic to mites.

The formulation used may influence not only the initial deposits but also the rates at which surface deposits weather. Dioxathion applied as an emulsifiable formulation resulted in a typical straight-line degradation curve after the first day, whereas residues resulting from a wettable powder formulation (Fig. 8A, line 2) weathered rapidly for 10 to 15 days. The deposition of dioxathion on navel oranges at each dosage of the EC formulation applied was about 2.5 times that deposited on lemons (compare Fig. 8A, line 1 and Fig. 8A, line 3); however, WP applications resulted in a higher deposit and longer effective persistence on lemons than on oranges (Fig. 8B). Two and one-half times as much residue was required on navel oranges as was needed on lemons to obtain equivalent mite mortality when the EC formulation was applied.

The amount of residue deposited by a WP formulation of carbophenothion on oranges and on lemons was nearly the same. Because lemons grew during the weathering (degradation) period there was an apparent more rapid decrease in actual and effective residue on lemons

Table IV. *Non-correlation among initial deposits, half-lives, and ultimate locations of residues in either lemons or oranges*

Acaricide	Avg. initial deposit [a] (p.p.m. on and in rind)		Half-life (days) [a]				Location of ultimate residue [b]
			Degradation		Persistence		
	Lemons	Oranges	Lemons	Oranges	Lemons	Oranges	
Neotran	16	15	~12	~5	30	39	oil sacs
EPN	–	~10	–	~5	80	50	aq. cells
Tetram	–	~5 [c]	– [d]	– [d]	–	~40 [c]	aq. cells
Ovex	7	–	12	–	10–30	10–25	oil sacs
Aramite	5	–	5	–	~13	~30	aq. cells
Chlorobenzilate	15	–	~25	–	~66	~66	wax
Carbophenothion	11	8	~10	–	~22	~42	oil sacs
Dioxathion	6	3	4	3	~70	>100	oil sacs
Dicofol	8	6	? [e]	? [e]	120–150	170–350	oil sacs
Tetradifon	4	7	15	–	~79	~130	wax
OW-9 compounds	–	30	–	~20	–	85–130	oil sacs
Morestan	4	7	–	4	~37	~36	aq. cells
Omite	–	22	–	~20	–	~75	oil sacs
Azodrin [f]	–	~7	–	5	–	12–14	aq. cells

[a] From Gunther (1969) and sources therein. Actual values for 0.5 lb. actual/ 100 gallons (direct or by extrapolation when necessary).
[b] Long-lasting biological effectiveness seems to be associated with long residence in the wax "layers" or in the aqueous epidermal and palisade cells.
[c] From present report.
[d] Systemic material.
[e] Not possible to estimate from original data.
[f] Unpublished data.

than on oranges (Fig. 7A and 7B). The amount of toxicant in the rind required to kill mites was slightly less on lemons than oranges when the WP formulation was applied, but less mortality resulted from equivalent amounts of residues when the EC formulation was applied (Fig. 7C).

Residues on leaves and fruit were analyzed following application of the acaricide OW-9. Deposits on leaves and fruit can be compared only on an area basis. On this basis deposits were about the same, but weathering was most rapid on the leaves. Degradation occurred at about the same rate on both surfaces; however, toxicity to the mites was longer on leaves than on fruit (Fig. 13B), as equivalent amounts of residue produced more mite mortality on fruit than on leaves.

Prevailing temperatures apparently influence the rate residues migrate into the rind. Warm summer temperatures decreased the duration of effective persistence of Neotran residues (Fig. 1) and no doubt of residues of other acaricides that generally migrate to the oil sacs. Although the half-life of the residues in the rind may not be influenced by weather, the location and thus the availability of the residues to the mites appears to be greatly influenced by weather variations.

Data for initial deposits, half-life values for both degradation and persistence curves, and the ultimate locations of the persisting residues of the parent compounds are collated in Table IV. Among these acaricides, EPN, Tetram, Aramite, Morestan, and Azodrin are long-lasting in effectiveness, demonstrating a good correlation with location of the ultimate residues in the epidermal and palisade cells. Chlorobenzilate and tetradifon residues also show long effectiveness, but they are apparently localized in the wax "layers" for long periods, moving very slowly into the subcuticular areas of the rind. There are no obvious correlations among biological effectiveness, initial deposit, and degradation or persistence behavior by chemical analyses, except that those acaricides with the shortest degradation half-lives seem to end up largely in the epidermal and palisade cells (dioxathion is the only exception).

Furthermore, at equivalent dosages (0.5 lb./100 gallons, extrapolated when necessary), there are no correlations between initial deposits on lemons vs. oranges (see also GUNTHER 1969), or between initial deposits and either type of half-life (see also GUNTHER and BLINN 1955, GUNTHER 1969). Correlations between chemical structures of these acaricides and the ultimate locations of their residues (oil sacs vs. aqueous tissues) are not apparent.

Summary

Following is a summary of the probable migration of acaricide residues in citrus fruits as evidenced by the availability of the residues to citrus red mite, either by contact or by feeding, as correlated with persisting residues established by chemical analysis:

Neotran. Effective residue duration is strongly influenced by temperature; also a poor correlation exists between residues found and their toxicity to mites, indicating that the residues concentrate in the oil sacs.

EPN. Residues rapidly penetrate into epidermal and palisade cells where they remain available to mites.

Tetram. Residues slowly penetrate into epidermal and palisade cells where they remain available to mites.

Ovex. Residues slowly penetrate into the rind and concentrate in the oil sacs after which they are no longer toxic to mite eggs and larvae.

Aramite. Residues slowly penetrate the rind and concentrate in epidermal palisade cells and remain there available to mites.

Chlorobenzilate. Residues remain on the surface or in the wax "layer" for long periods. Residues above 10 p.p.m. were toxic to citrus red mite; lower residues were toxic to eriophyid mites.

Carbophenothion. Residues penetrate slowly, ultimately migrate to the oil sacs where they are not available to mites.

Dioxathion. Residues rapidly penetrate into the rind and concentrate in the oil sacs where they are not available to mites; formulations and citrus variety influence residues during the time interval in which they are toxic to mites.

Dicofol. Residues slowly penetrate into the rind where they concentrate in the oil sacs and thus are not available to mites; also, little correlation exists between amounts of residue and their toxicities to mites.

Tedion. Residues are rapidly absorbed by the wax "layer" of the fruit where they remain toxic to mites; they can only be removed by solvents or strong detergents which also remove some of the wax.

OW-9. Residues rapidly penetrate the rind where they are concentrated in the oil sacs and thus are not available to mites; little correlation exists between amounts of residue and their toxicities to mites.

Morestan. Residues remain on the fruit surface for a considerable period as evidenced by their ready attenuation by rains or by water washing and by their long-term toxicities to mites.

Omite. Residues migrate into the rind and there concentrate in the oil sacs where they are unavailable to mites.

Azodrin. Residues rapidly migrate into the epidermal and palisade cells where they remain available to mites.

It is clear that under the described circumstances there is no general correlation between residue persistence and biological effectiveness of citrus acaricides. Acaricides which are soluble in citrus oil penetrate rapidly and become unavailable to mites either by contact or by feeding, whereas more water-soluble acaricides may persist longer on fruit and leaf surfaces or penetrate into epidermal and palisade cells where they may be available to feeding mites. Penetrated acaricides show markedly different degrees of both degradation and persistence behavior on and in mature citrus fruits. The residues of a few acaricides which are not soluble in water or citrus oils either remain on the surface (*i.e.*, Morestan) or are adsorbed on the wax "layer" (*i.e.*, tetradifon) where they can be contacted by the mites.

Résumé *

Résidus d'acaricides sur le feuillage et les fruits des citrus et leur importance biologique

On donne ci-après un résumé de la migration probable des résidus d'acaricides dans les agrumes, d'après l'utilisation des résidus par l'araignée rouge des citrus, par contact ou ingestion, et la relation avec les résidus persistants établis par l'analyse chimique.

---

* Traduit par S. DORMAL-VAN DEN BRUEL.

Neotran. La persistance des résidus effectifs est fortement influencée par la température; il n'existe qu'une faible relation entre les résidus trouvés et leur toxicité pour les acariens, ce qui indique que les résidus se concentrent dans les glandes à huile essentielle.

EPN. Les résidus pénètrent rapidement dans les cellules épidermiques et palissadiques où ils restent disponibles pour les acariens.

Tetram. Les résidus pénètrent lentement dans les cellules épidermiques et palissadiques où ils restent disponibles pour les acariens.

Ovex. Les résidus pénètrent lentement dans l'écorce et se concentrent dans les glandes à huile essentielle où ils perdent leur toxicité pour les oeufs et les larves d'acariens.

Aramite. Les résidus pénètrent lentement dans l'écorce et se concentrent dans les cellules épidermiques et palissadiques où ils restent disponibles pour les acariens.

Chlorobenzilate. Les résidus se maintiennent durant de longues périodes en surface ou dans la "couche" cireuse. Les résidus supérieurs à 10 ppm étaient toxiques pour l'araignée rouge des citrus; les résidus moins importants toxiques pour les acariens ériophyides.

Carbophénothion. Les résidus pénètrent lentement; ils migrent finalement dans les glandes à huile essentielle où ils ne sont plus disponibles pour les acariens.

Dioxathion. Les résidus pénètrent rapidement dans l'écorce et se concentrent dans les glandes à huile essentielle où ils ne sont plus disponibles pour les acariens; les formulations et la variété de citrus influencent les résidus durant la période au cours de laquelle ils sont toxiques pour les acariens.

Dicofol. Les résidus pénètrent lentement dans l'écorce où ils se concentrent dans les glandes à huile essentielle et ne sont donc plus disponibles pour les acariens. Il n'existe qu'une faible relation entre les quantités de résidus et leur toxicité pour les acariens.

Tedion. Les résidus sont rapidement absorbés par la "couche" cireuse du fruit où ils restent toxiques pour les acariens; ils ne peuvent être éliminés que par des solvants ou de forts détergents qui éliminent également une part de la cire.

OW-9. Les résidus pénètrent rapidement dans l'écorce où ils se concentrent dans les glandes à huile essentielle et sont donc disponibles pour les acariens; il n'existe qu'une faible relation entre les quantités de résidus et leur toxicité pour les acariens.

Morestan. Les résidus se maintiennent durant une longue période en surface des fruits; ceci est mis en évidence par leur prompte élimination par les pluies et par lavage à l'eau et par leur toxicité à long terme pour les acariens.

Omite. Les résidus migrent dans l'écorce et se concentrent dans les glandes à huile essentielle où ils ne sont plus disponibles pour les acariens.

Azodrine. Les résidus migrent rapidement dans les cellules épidermiques et palissadiques où ils restent disponibles pour les acariens.

Il est clair que, dans les conditions décrites, il n'y a pas de relation générale entre la persistance des résidus et l'efficacité biologique des acaricides des citrus. Les acaricides solubles dans les huiles essentielles des agrumes pénètrent rapidement et ne sont plus disponibles au contact ou à l'ingestion par les acariens, tandis que les acaricides plus solubles dans l'eau peuvent persister plus longtemps sur le fruit et les surfaces foliaires ou pénétrer dans les cellules épidermiques et palissadiques où ils peuvent être disponibles à l'ingestion par les acariens. Les acaricides qui pénètrent montrent nettement des degrés différents dans le comportement de la dégradation et de la persistance sur et dans les agrumes à maturité. Les résidus de quelques acaricides qui ne sont solubles ni dans l'eau, ni dans les huiles essentielles restent en surface (par ex. Morestan) ou sont adsorbés sur la "couche" cireuse (par ex. tetradifon) où ils peuvent subir le contact des acariens.

Zusammenfassung *

Rückstände von Akariziden auf Zitrus-Blattwerk und-Früchten und ihre biologische Bedeutung

Das Folgende ist eine Zusammenfassung der möglichen Wanderung akarizider Rückstände in Zitrus-Früchten. Die Verfügbarkeit der Rückstände für die rote Zitrus-Milbe—entweder durch Kontakt oder Fütterung—und deren Wechselbeziehung zu persistenten Rückständen, die durch chemische Analyse bestimmt wurden, wird unter Beweis gestellt:

Neotran. Wirksame Rückstandsdauer wird durch Temperatur stark beeinflusst; auch besteht eine geringe Wechselbeziehung zwischen gefundenen Rückständen und ihrer Toxizität gegenüber Milben, was darauf hinweist, dass sich die Rückstände in den Ölsäcken konzentrieren.

EPN. Rückstände dringen schnell in die Epidermis- und Pallisaden-Zellen ein, wo sie für die Milben erreichbar bleiben.

Tetram. Rückstände dringen langsam in die Epidermis- und Pallisaden-Zellen ein, wo sie für die Milben erreichbar bleiben.

Ovex. Rückstände dringen langsam in die Rinde ein und konzentrieren sich in den Ölsäcken; danach sind sie für Milbeneier und Larven nicht länger toxisch.

Aramite. Rückstände dringen langsam in die Rinde ein und konzentrieren sich in Epidermis- und Pallisaden-Zellen; sie bleiben dort für die Milben erreichbar.

Chlorbenzilat. Rückstände bleiben für lange Zeit an der Oberfläche oder in der Wachs—"Schicht". Rückstände über 10 p.p.m. waren für die rote Zitrus-Milbe toxisch; niedrigere Rückstände waren toxisch für Eriophyid-Milben.

---

* Übersetzt von M. Dusch.

Carbophenothion. Rückstände dringen langsam ein und wandern schliesslich zu den Ölsäcken, wo sie für die Milben nicht erreichbar sind.

Dioxathion. Rückstände dringen schnell in die Rinde ein und konzentrieren sich in den Ölsäcken, wo sie für die Milben nicht erreichbar sind; Formulierungen und Zitrus-Verschiedenartigkeit beeinflussen die Rückstände während der Zeitspanne in der sie für Milben toxisch sind.

Dicofol. Rückstände dringen langsam in die Rinde ein, wo sie sich in den Ölsäcken konzentrieren und so für die Milben nicht erreichbar sind; auch besteht wenig Wechselbeziehung zwischen Rückstandsmengen und ihren Toxizitäten gegenüber Milben.

Tedion. Rückstände werden schnell von der Wachs—"Schicht" der Frucht absorbiert, wo sie für Milben toxisch bleiben; sie können nur durch Lösungsmittel oder starke Waschmittel entfernt werden, die auch etwas von dem Wachs entfernen.

OW-9. Rückstände dringen langsam in die Rinde ein, wo sie in den Ölsäcken konzentriert werden und so für die Milben nicht erreichbar sind; wenig Wechselbeziehung besteht zwischen Rückstandsmengen und ihren Toxizitäten gegenüber Milben.

Morestan. Rückstände bleiben für einen beträchtlichen Zeitraum auf der Fruchtoberfläche, wie durch ihre schnelle Abschwächung durch Regen oder Waschen mit Wasser und durch ihre langen Toxizitäten gegenüber Milben bewiesen wurde.

Omite. Rückstände wandern in die Rinde und konzentrieren sich dort in den Ölsäcken, wo sie für Milben nicht erreichbar sind.

Azodrin. Rückstände dringen schnell in die Epidermis- und Pallisaden-Zellen ein, wo sie für Milben erreichbar sind.

Es ist klar, dass unter den beschriebenen Umständen keine allgemeine Wechselbeziehung zwischen Rückstandshartnäckigkeit und biologischer Wirksamkeit von Zitrus-Akariziden besteht. Akarizide, die in Zitrus-Öl löslich sind, dringen schnell ein und werden für Milben—entweder durch Kontakt oder durch Fütterung—unerreichbar, während wasserlöslichere Akarizide länger auf Frucht- und Blattoberflächen bestehen können oder in Epidermis- und Pallisaden-Zellen eindringen, wo sie für fressende Milben erreichbar sind. Eingedrungene Akarizide zeigen deutlich verschiedene Grade sowohl in Degradations- als auch in Hartnäckigkeitsverhalten auf und in reifen Zitrusfrüchten. Die Rückstände von den wenigen Akariziden, die nicht in Wasser oder Zitrus-Ölen löslich sind, bleiben auf der Oberfläche (z.B. Morestan) oder werden auf der Wachs—"Schicht" adsorbiert (z.B. Tetradifon), wo sie mit den Milben in Berührung kommen können.

References

BLINN, R. C., R. W. DORNER, J. H. BARKLEY, L. R. JEPPSON, F. A. GUNTHER, and C. C. CASSIL: Locale of aged TEDION residues on citrus fruits. J. Econ. Entomol. **52**, 723 (1959).

Cassil, C. C., and O. H. Fulmer: Persistence of TEDION residues on fruits. J. Agr. Food Chem. 6, 908 (1958).

Crafts, A. S., and C. L. Foy: The chemical and physical nature of plant surfaces in relation to the use of pesticides and to their residues. Residue Reviews 1, 112 (1962).

Ebeling, W.: Analysis of the basic processes involved in the deposition, degradation, persistence, and effectiveness of pesticides. Residue Reviews 3, 35 (1963).

Gage, J. C.: Residue determination by cholinesterase inhibition analysis. Adv. Pest Control Research 4, 194 (1961).

Gunther, F. A.: Insecticide residues in California citrus fruits and products. Residue Reviews 28, 1 (1969).

———, and R. C. Blinn: Analysis of insecticides and acaricides. New York-London: Interscience (1955).

——— ———, L. R. Jeppson, J. H. Barkley, G. J. Frisone, and R. D. Garmus: Field persistence of the acaricide 4,4'-dichloro-alpha-(trichloromethyl) benzhydrol (FW-293) on and in mature lemons and oranges. J. Agr. Food Chem. 5, 595 (1957).

———, M. J. Kolbezen, and J. H. Barkley: Microestimation of 2-(p-tert-butylphenoxy) isopropyl-2-chloroethyl sulfite residues. Anal. Chem. 23, 1835 (1951).

———, G. E. Carman, L. R. Jeppson, J. H. Barkley, and R. C. Blinn: Residual behavior of S-(p-chlorophenylthio) methyl O,O-diethyl phosphorodithioate (Trithion) on and in mature lemons and oranges. J. Agr. Food Chem. 7, 28 (1959).

———, and L. R. Jeppson: Residues of p-chlorophenyl p-chlorobenzenesulfonate (Compound K-6451) on and in lemons and oranges. J. Econ. Entomol. 47, 1027 (1954).

——— ——— Data on file at Univ. Calif., Riverside (1969).

——— ———, J. H. Barkley, L. M. Elliott, and R. C. Blinn: Persistence of residues of 2,3-p-dioxathanedithiol S,S-bis(O,O-diethyl phosphorodithioate) as an acaricide on and in mature lemons and oranges. J. Agr. Food Chem. 6, 210 (1958).

———, and W. E. Westlake: Studies of organic insecticide residues in California citrus fruits and products. Proc. First Internat. Citrus Symposium 2, 1053 (1959).

——— ———, and P. S. Jaglan: Reported solubilities of 738 pesticide chemicals in water. Residue Reviews 20, 1 (1968).

Jeppson, L. R.: Data on file at Univ. Calif., Riverside (1950).

——— Bis-(p-chlorophenoxy) methane in relation to the control of citrus red mite and other mites injurious to citrus in California. J. Econ. Entomol. 44, 326 (1951).

———, M. J. Jesser, and J. O. Complin: Control of mites on citrus with chlorobenzilate. J. Econ. Entomol. 48, 375 (1955).

——— ——— ——— Effectiveness of two new phosphate insecticides for control of mites injurious to citrus in California. J. Econ. Entomol. 50, 307 (1957).

——— ——— ——— Resistance of the citrus red mite to organic phosphates in California. J. Econ. Entomol. 51, 232 (1958).

——— ——— ——— Response of citrus mites to selected quinoxaline cyclic di-on trithiocarbonates. J. Econ. Entomol. 60, 994 (1967).

———, W. E. Westlake, and F. A. Gunther: Toxicity, control and residue studies with do-14 [2-(p-tert-butylphenoxy)cyclohexyl 2-propyl sulfite] as an acaricide against the citrus red mite. J. Econ. Entomol. 62, 531 (1969).

Linskens, H. F., W. Heinen, and A. L. Stoffers: Cuticula of leaves and the residue problem. Residue Reviews 8, 136 (1965).

Metcalf, R. L., R. B. March, and T. R. Fukuto: The behavior of SYSTOX isomers in bean and citrus plants. J. Econ. Entomol. 47, 780 (1954).

Subject Index

ABC persistence curves 109, 110
AC 12008 metabolism by algae, bacteria, and yeasts 31 ff.
Acaricide residues, biological significance (see also specific compounds) 101 ff.
Acaricides, qualities necessary to evaluate 102
Acceptable daily intake 76, 78
Acetaldehyde, minimum interval, Poland 9
Agriculture in Poland 2 ff.
Aldrin, effect on oysters and mussels 23
——— effect on plankton 19
——— from dieldrin in soil 31
——— metabolism by microorganisms 29
——— minimum interval, Poland 8
——— tolerance, Poland 12
Alewives, DDT in 25
Algae (see specific compounds, effects on) 16, 18, 21, 22, 24, 25, 31, 33
Almonds 82
Ametryne photolysis 64
Amiben photochemistry 57
——— photolysis 52
Amidithion in Czechoslovakia 77
Apples 10, 11, 77, 80, 81
Aquatic environment, entry of pesticides into 16, 17
——— environment, pesticides detected in 17
——— environment, sources of pesticides in 17
——— microorganisms and pesticides 15 ff.
——— microorganisms, definition 15, 16
——— microorganisms, toxicity of pesticides to 17 ff.
Aramite 128, 129, 131
——— residues, biological significance 111 ff.
——— residues, half-life 130
——— residues, location in citrus fruit 130

——— residues on and in lemons and oranges 111, 112
——— residues vs. mite toxicity 112
ASP-51, effect on plankton 19
Atrazine metabolism by bacteria 33 ff.
——— photolysis 64
Azinphos ethyl, effect on Daphnia 22
——— methyl, effect on bacteria 22
Azodrin 128, 131
——— residues and mite toxicity 127, 128
——— residues, biological significance 126 ff.
——— residues, half-life 130
——— residues, location in citrus fruit 130
——— residues on and in oranges 128

Bacon, MGK 264 in 88–90
——— piperonyl butoxide in 88–90
——— pyrethrins in 88–90
Barban, minimum interval, Poland 9
Barley 81
Bayer 37344, effect on plankton 19
Bay leaves 82
Baytex, effect on plankton 20
Beef tallow, DDT in 80
Beets 2, 3, 10, 77
Berries, see specific kinds
BHC, effect on bacteria 22, 23
——— in Poland, prohibition 6
——— minimum interval, Poland 8
——— persistence in soils 27, 35
——— tolerance, Poland 12
Biological significance of acaricide residues (see also specific compounds) 101 ff.
Biphenyl, see Diphenyl
Bond dissociation energies for solvents 53, 54
Brazil nuts, MGK 264 in 88–90
——— nuts, piperonyl butoxide in 88, 90
——— nuts, pyrethrins in 88, 90
Bread 81
Bromacil photolysis 68
Bromophenol photolysis 52